CIRIA C538 London, 2000

A review of testing for moisture in building elements

Michael J Dill

sharing knowledge ▪ building best practice

6 Storey's Gate, Westminster, London SW1P 3AU
TELEPHONE 0207 222 8891 FAX 0207 222 1708
EMAIL enquiries@ciria.org.uk
WEBSITE www.ciria.org.uk

Summary

This guide provides information to enable the reader to gain a basic understanding of moisture in materials and building elements and the need for testing. It gives details on the range of techniques available for determining moisture presence in building elements and which test method(s) are the most appropriate to use to assess a particular situation. The guide stresses the importance of clear understanding of these issues when specifying tests to be used to ensure that quality control and long-term monitoring requirements are achieved. The guide also provides independent guidance to facilitate the selection of the appropriate technique(s) so that moisture problems can be detected early and defect investigation can be efficiently and effectively carried out.

A review of testing for moisture in building elements

Dill, M J

Construction Industry Research and Information Association

CIRIA Publication C538 CIRIA 2000 ISBN 0 86017 538 3

Keywords

Automated long-term monitoring, built-in water, calcium carbide moisture meter, dampness, defect investigation, earth leakage, electrical capacitance, electrical resistance, humidity sensors, hygroscopic, impedance, infrared, moisture movement, microwave, moisture content, moisture, nuclear magnetic resonance, oven drying, quality control, radar, relative humidity, resistivity, rising damp, thermographic inspection, thermography, ultrasonics, ultrasound, vapour pressure

Reader interest	**Classification**	
Designers, contractors, surveyors and maintenance personnel	AVAILABILITY	Unrestricted
	CONTENT	Guidance document
	STATUS	Committee-guided
	USER	Designers, contractors, surveyors, maintenance personnel

Published by CIRIA, 6 Storey's Gate, Westminster, London SW1P 3AU.

Acknowledgements

This guide was produced as a result of CIRIA Research Project 581, "A review of test methods for measuring moisture in building elements". The authors listed below provided the reviews of the different methods given in Section 5. The research contractor, Laing Technology Group Ltd, was responsible for co-ordinating and editing Section 5 and for producing the remainder of the guide. The principal contributors were Gordon Sparksman and Michael Dill.

The project was developed and managed by Ann Alderson, research manager at CIRIA, with advice and guidance from a project steering group, whose support and valuable contributions are gratefully acknowledged.

The members of the steering group were:

Mr Guy Hammersley (chairman)	Building Research Establishment
Mr David Clifton	Munters MCS
Dr Chris Coggins	Institute of Wood Preserving and Damp-proofing
Mr David Cridland	Vaisala (UK) Ltd
Mr Peter Dickinson	Premises Consultants *representing* *The Royal Institute of Chartered Surveyors*
Mr Sean Fallon	Tramex Ltd
Mr Robin Farley/Mr Mike Hibbert	Rotronics Instruments (UK) Ltd
Mr Philip Leach	Protimeter plc
Dr Maurice Levitt	PR Consultant *representing the Concrete Society*
Mr Robert Mabbutt	Laybond Products Ltd
Miss Sarah McKenna	Ove Arup & Partners
Mr Peter Woodhead	DETR.

Corresponding member:

Mr Jim Hooker	British Flat Roofing Council.

The research leading to this publication was funded by:

The Department of the Environment, Transport and the Regions (DETR)
Institute of Wood Preserving and Damp-proofing (IWPD)
Laybond Products Ltd
Munters MCS
Protimeter plc
Rotronics Instruments (UK) Ltd
Tramex Ltd
Vaisala (UK) Ltd.

The authors of the reviews in Section 5 were:

Dr Kemel Ahmet	Department of Science Technology and Design, University of Luton
Mr Jerry Assenheim	Physical and Electronic Laboratories Ltd
Mr George Ballard	GBG Ltd
Mr Sam Dods	Aperio Ltd
Mr Chris Howard	School of the Built Environment, Liverpool John Moores University
Mr Tim Hutton	Hutton & Rostron
Mr David Mostyn	David Mostyn Consultancy Ltd
Mr Steve Pringle	Sandberg Consulting Engineers
Mr Keith Roberts	Glanville Consultants
Mr David Rolfe	Control Testing Equipment Ltd
Mr Chris Sanders	BRE Scottish Laboratory
Mr Andrew Tee	Quest Technical Services Ltd
Mr Steve Thornton	Thornton Consulting Group
Professor Jonathan Wood	Structural Studies and Design Ltd

CIRIA is grateful to them and to Laing Technology Group Ltd for their in-kind contributions to the project.

Contents

LIST OF FIGURES

LIST OF TABLES

Glossary

absorbed water — Water (vapour) penetrating into the body of a material by **absorption**.

absorption — Penetration of a substance into the body of another.

accuracy — The smallest increment that can be measured with confidence and repeatability, always larger value than **sensitivity**.

adsorbed water — Water (vapour) taken up in at the surface of a material by **adsorption**.

adsorption — The taking up of one substance at the surface of another.

blister — An elevation of the surface of an adherent, its boundaries may be indefinitely outlined, and it may have burst and become flattened. Note: a blister may be caused by insufficient adhesive; inadequate curing times, temperature or pressure; or trapped air, water or solvent vapour. See Section 2.1 *Blistering.*

bloom — In painting, a dull film that develops on a gloss surface, eg after painting in humid conditions. Also the discoloration of material surface.

bound water — Water that is bound chemically (by crystallisation and hydration).

built-in water — Residual moisture from the building process normally associated with wet trades, poor storage conditions or inadequate protection during construction. See Section 2.2 *Built-in water.*

capacitance — See **electrical capacitance**.

capillarity — The action by which the surface of a liquid (where it is in contact with a solid) is elevated or depressed, depending upon the relative attraction of the molecules of the liquid for each other and for those of the solid. See Section 2.3 *Capillary action.*

capillary water — Water (liquid and/or vapour) that is contained in connected voids or capillaries within the body of the material.

chemically bound water — See **water of crystallisation** and **water of hydration**.

condensation — The change of state from water vapour to liquid water (can be considered the opposite of evaporation). See Section 2.2 *Condensation.*

corrosion — Commonly applied to the oxidisation of metals by an electrochemical mechanism. See Section 2.1 *Corrosion.*

data logger — A programmable device that automatically takes the output from a number of **sensors** at regular time intervals, converts the outputs to the measurement units and stores the data. The device can be directly interrogated or interfaced to a local or remote PC to display the current or historical data. See Section 6 *Automated long-term monitoring.*

decay — The decomposition of wood substance by fungi. Section 2.1 *Decay.*

delamination — Delamination is the breakdown of the bond at the interface between two different materials or two layers of the same material, or a fracture that forms within a material and is near parallel to the surface. See Section 2.1 *Delamination.*

dew-point temperature — The temperature at which water vapour starts to condense in cooling air at the existing atmospheric pressure and vapour content. See Section 2.2 *Condensation.*

dry-bulb temperature	The measured temperature of air as indicated by an ordinary thermometer (ºC). When referring to air temperature, dry-bulb temperature is always used unless otherwise stated.
efflorescence	White crystalline deposits formed on a material, following evaporation, from soluble salts transported to the surface by water. See Section 2.1 *Efflorescence.*
electrical capacitance	The total electric charge on a conducting body divided by its potential (C=Q/V). The measurement unit is the farad. See Section 5.2 *Electrical capacitance meter.*
electrical resistance	The ability of a conductor to oppose the flow of electric current. The unit of measurement is the ohm. See Section 5.2 *Electrical resistance meter.*
electrical resistivity	The specific property of a conductor, which gives the resistance in terms of its dimensions eg kohm cm.
equilibrium moisture content	The value of **moisture content** that remains constant for given conditions of temperature and **relative humidity**.
free moisture content	That portion of the total moisture in a material that is relatively easily driven off, ie by oven drying but would not include chemically bound water. Also termed **moisture content**.
hygroscopic	The readiness of some materials to absorb moisture, sometimes directly from the air.
hygroscopic moisture	The amount of moisture which is absorbed from the atmosphere. This is dependent on the nature and composition of the material and the ambient moisture content of the atmosphere of the space in which the element is located. High relative humidity will be linked with high hygroscopic moisture absorption in some materials.
hygroscopic moisture content (HMC)	HMC is the **moisture content** of a material after it has reached equilibrium in an atmosphere at a given **relative humidity**. See Section 5.9 *Oven drying method.*
impedance	Opposition to the flow of alternating current. It can include **electrical resistance**. See Section 5.4 *Electrical resistance meter.*
infrared radiation	A form of electromagnetic radiation with wavelengths slightly longer than red light. It is not visible but can be felt as heat eg radiation from the sun and can be detected by photographic film and other devices. See Section 5.11 *Thermographic inspection.*
leaching	The depletion of soluble products from within a material caused by the migration of water through a material. See Section 2.1 *Leaching.*
moisture content (MC)	The moisture content of a material is expressed as a percentage value based on the original wet or final dry measurement of weight or volume. The basis of the measurement should be defined eg free MC = X per cent by dry weight. See Section 3.3 *Moisture measurements* and Section 5.9 *Oven drying method.*
osmosis	The diffusion of a solvent through a semi-permeable membrane into a more concentrated solution tending to equalise the concentrations. See Section 2.2 *Osmosis.*
oven dry	Material dried to a constant weight usually in a ventilated oven at 102–105ºC. See Section 5.9 *Oven drying method.*
permeance	An index of material's resistance to water vapour transmission.
pH	A simple chemical scale that expresses the degree of acidity or alkalinity of a solution. The pH scale runs from 0 to 14, with 7 being the neutral point. Numbers below 7 indicate acidity, and those above 7 indicate alkalinity.
pore water	Water (liquid and/or vapour) that is contained discrete voids within the body of the material.
pyschrometer	An instrument for measuring the amount of water vapour in the atmosphere.

quality assurance	All the planed and systematic activities necessary to provide confidence that a product or service will satisfy given requirements for quality (planned work carried out to a plan).
quality control	The use of operational techniques and activities that sustain the quality of product and services to specified requirements (checking against known criteria).
relative humidity	The ratio of the amount of water vapour present in the air to that which the air would hold at saturation at the same temperature. It is usually considered on the basis of the weight of the vapour, but, for accuracy, should be considered on the basis of **vapour pressure**.
resistance	See **electrical resistance**.
saturation	A condition existing when the air contains as much water vapour as it can hold at a specific temperature and pressure. It also represents 100 per cent relative humidity. At saturation, **dry-bulb** and **dew point temperature** are the same readings.
sensitivity	The smallest increment that can be detected with confidence but not necessarily measured with accuracy.
sensor	The active part of an instrument that reacts to the parameter being measured. May be loosely termed probe or transmitter. See Section 6 *Automated long-term monitoring*.
specific humidity	The mass of water vapour in a unit mass of moist air, ie percentage moisture content by wet weight.
vapour pressure	The pressure exerted by water vapour in a space at its corresponding **dew-point temperature** (that part of the total pressure contributed by water vapour).
vapour pressure diffusion	The movement of moisture vapour through the building fabric in order to reach equilibrium **vapour pressure** on both sides of the fabric. See Section 2.3 *Vapour pressure diffusion*.
water of crystallisation	The fixed proportion of water that is chemically combined with certain substances when they are crystals.
water of hydration	The proportion of water that is combined with certain substances as result of a chemical reaction between water and the material eg cement. During the hydration process cement paste sets and hardens while heat is given off.
wet-bulb temperature	The temperature indicated by a thermometer bulb that is encased by a piece of muslin saturated with water, ie influenced by the rate of evaporation of water to the air. It is used in conjunction with **dry-bulb temperature** to derive **relative humidity**.

Abbreviations

AAR	alkaline aggregate reaction
ASTM	American Society for Testing and Materials
BCA	British Cement Association
BRE	Building Research Establishment
BMS	building management system
BWPDA	British Wood Preserving and Damp-proofing Association *formerly* British Wood Preserving Association (BWPA)
CDM	Construction (Design and Management)
CE	Comité Européen
COSHH	Control of Substances Hazardous to Health
CMP	common mid point
CW	continuous wave
dpc	damp proof course
dpm	damp proof membrane
EP	ethylene propylene
EPDM	ethylene propylene diene monomer
ERH	equilibrium relative humidity
GRP	glass-reinforced plastic
HMC	hygroscopic moisture content
ISO	International Organisation for Standardisation
LED	light-emitting diode
MC	moisture content
NBS	National Building Specification
NDT	non-destructive testing
NIST	National Institute of Standards and Technology
NMR	nuclear magnetic resonance
NPL	National Physical Laboratory
PC	personal computer
pvc	polyvinyl chloride (thermo-plastic)
PPE	personal protective equipment
QC	quality control
RF	radio frequency
rh	relative humidity
TRADA	Timber Research and Development Association
UKAS	United Kingdom Accreditation Service *formerly* National Measurement Accreditation Service (NAMAS)
WME	wood moisture equivalent

1 Introduction

1.1 THE NEED FOR GUIDANCE

Designers, contractors, surveyors and maintenance personnel are often not aware of the range of techniques available for determining moisture presence in building elements and which method(s) are the most appropriate to assess a particular situation. There is also generally little published guidance on the test methods apart from manufacturer or operator's literature. This report aims to fill that gap by providing information on the range of techniques that can be employed to ensure that quality control and long-term monitoring requirements are achieved. The report also provides independent guidance to facilitate the selection of the appropriate technique(s) so that moisture problems can be detected early and defect investigation can be efficiently and effectively carried out.

These are the three basic situations where moisture testing may be needed:

- quality control
- long-term monitoring
- defect investigation.

1.2 BACKGROUND

Moisture is a primary cause of a large percentage of problems in buildings, both during and after construction. Moisture will always be present in, even essential to all buildings; it needs to be controlled to maintain a level within acceptable limits during both construction and the operational life. Moisture measurements provide essential quantitative data to demonstrate adequate control. Control is achieved through:

- building design eg adequate ventilation to minimise condensation problems
- correct construction eg adequate drying times for wet trade
- adequate maintenance eg to prevent rain penetration at joints/seals.

The rectification of the damage caused by both low and high moisture conditions within building elements can be extremely difficult, disruptive and costly to repair. Too low a moisture content can result in shrinkage, cracking and splitting in timber with potential loss of structural integrity. Excess moisture can:

- increase atmospheric moisture levels leading to increased risk of condensation problems in other parts of the building
- cause damage to materials that may compromise structural integrity
- increase the thermal conductivity of some insulating materials, which results in greater energy loss
- permit the migration of salts and other chemicals that can cause secondary damage
- lead to degradation of the building environment such that the conditions become unfit for habitation as defined under the Environmental Protection Act.

With increased emphasis being placed on quality assurance and long-term performance, some major clients require certification for building elements. This requires measurements, quality control, to demonstrate the level of built-in moisture and to establish the effectiveness of waterproof membranes, seals and sealants; ie to demonstrate the minimisation of potential defects.

1.3 SCOPE

The guide has been written to provide independent, authoritative guidance for designers, contractors, surveyors and maintenance personnel on different techniques available for detecting or measuring moisture in building elements. It is primarily a compilation of reviews of the different methods, and is not intended to be a definitive reference document on moisture in building elements, its causes, behaviour, diagnosis and remedies. It aims to provide information on the principles and characteristics of each method, appropriate use, advantages and limitations, means of operation, interpretation of the results and further key sources of reference for those requiring greater detail. It does not attempt to compare the different methods. The methods reviewed, in Section 5, by different authors, are:

- calcium carbide moisture meter
- electrical capacitance meter
- electrical earth leakage technique
- electrical resistance meter
- humidity sensors
- microwave moisture meter
- nuclear moisture gauge
- nuclear magnetic resonance
- oven drying method
- radar
- thermographic inspection
- ultrasonics in the detection of leaks.

Ancillary investigations and tests that may be required in conjunction with the methods reviewed to carry out a full investigation are mentioned but not covered in detail. These are identified in the text by bracketing the name and include:

- air pressure tests
- breakout investigations
- borescope inspection
- chemical tests
- flood/spray/dye testing
- flow meter (to monitor water usage patterns)
- hygroscopicity tests
- leak detection tapes
- resistance probes/resistivity tests
- temperature sensors
- visual inspections.

Introductory sections provide a basic understanding of moisture in materials and building elements, and the need for testing. This includes comment on the measurement of moisture during construction (quality control), for long-term monitoring applications and the detection and diagnosis of defects. Finally guidance is given on selecting the appropriate method or methods for a particular situation. Case studies have been included to show where, when and how the benefits of using the test are achieved and to highlight common pitfalls.

It is essential that readers appreciate that this document is not a comprehensive guide to all the procedures involved in a quality control, long-term monitoring or defect diagnosis situation. **Simply reading this guide will not enable the reader to become an expert: that takes time and experience.**

While different authors wrote the reviews and contributed to other sections of the guide, the document has received peer review by the members of the Steering Group and other authors. The guide represents the consensus view of this group.

1.4 HOW TO USE THE GUIDE

There is a strong tendency for defect investigation to take precedence over quality control and long-term monitoring. However contributors to this guide recommend that it is imperative that not only should more attention be given to "doing it right first time" but that it should also be demonstrated. Hence throughout this guide quality control testing and long-term monitoring appear before defect investigation.

Section 2 provides a short introduction to the main parameters to be considered relating to the moisture effects on building elements. The descriptions are intended to inform and later the headings can be used as prompts/checklists/reminders.

Section 3 provides an introduction to testing for moisture with example applications for quality control, long-term monitoring and defect investigation.

Section 4 provides an introduction to the strategy for defect investigation, example applications and checklists. Test method selection tables are given at the end of this section. The tables are in the order recommended quality control, long-term monitoring and defect investigation.

Section 5 gives detailed guidance on the individual test methods, which are ordered alphabetically.

Section 6 gives detailed guidance on automated long-term monitoring.

Section 7 lists the identified standards, legislation, specifications, general references and a bibliography of further reading.

Table 1.1 indicates the likely relevance of the next three sections in the guide for the three main reasons for testing. If there is a temptation to jump directly to **only** those sections that are identified as essential, please at least read the example applications **Tables 4.1 and 4.2**. It is too easy to **jump** to the wrong conclusions!

Table 1.1 *Indicating the likely relevance of the next three sections in the measurement of moisture in building elements*

		Quality Control Long-term monitoring	Defect investigation
Section 2 – Understanding moisture			
2.1	Effects of moisture		
2.2	Sources of dampness		
2.3	Mechanisms of moisture movement		
2.4	Elements and environments		
2.5	Materials and interfaces		
Section 3 – Testing for moisture			
3.1	Objectives of testing		
3.2	Reasons for testing		
3.3	Moisture measurements		
3.4	Specifications and standards		
3.5	Health and Safety		
3.6	Training/certification		
3.7	A final word		
Section 4 - Selecting the appropriate test method			
4.1	Diagnostic strategy/procedures		
4.2	Example applications (Tables 4.1 and 4.2)		
4.3	Criteria for test selection, (Table 4.3)		
4.4	Test selection tables		
	Table 4.4		
	Table 4.5		
	Table 4.6		
	Table 4.7		

Legend

	Need to be aware
	Should understand
	Essential to have good understanding

2 Understanding moisture

2.1 EFFECTS OF MOISTURE

The following paragraphs give a short description of **moisture-** (liquid or vapour) related effects of damp conditions, defects, found in building materials/elements. In the investigation of these defects the source of moisture and mechanism for moisture movement need be identified. (See Section 2.2 *Sources of dampness*, Section 2.3 *Mechanisms of moisture movement* and Section 4.1 *Diagnostic strategy/procedures.)*

Blistering Blistering is usually caused by moisture migrating to the surface of a material below an impervious coating. It may also be caused by chemical damage, eg as the result of alkaline salts from freshly laid cementitious products attacking some plastics including floor coverings.

Corrosion Corrosion is the term commonly applied to the oxidation of metals by an electrochemical mechanism. This process requires the presence of air, water and a pH of less than 12 to form and maintain an electrical cell.

Decay Decay is usually associated with wood (organic materials) and is the decomposition/ rotting of the material in the presence of excess water (eg wet rot in wood at a moisture content of >25 per cent). Wet rot is a fungal form of decay that stops, ie no further decay, once the moisture source has been eliminated. (See *Fungal infestation/growth.*)

Delamination Delamination is the breakdown of the bond at the interface between two different materials or two layers of the same material or a fracture that forms within a material and is near parallel to the surface. Moisture is one agent capable of promoting delamination. Moisture can also result from delamination, eg detachment from silicone or other mastic in curtain walling elements or failure in expansion joints.

Efflorescence Efflorescence is the formation of white crystalline deposits on the surface of masonry as the result of loss of water from salts on exposure to air. (See *Leaching.*)

Freeze/thaw damage Freeze/thaw damage is caused by restrained expansion of water when it freezes within voids. This causes stresses that exceed the material strength and leads to progressive cracking; the cracks being refilled with water before the next freeze/thaw cycle.

Fungal infestation/ growth Moisture can support/sustain fungal infestation/growth that develop within organic materials as fluffy, woolly or mushroom growths and that can spread by releasing air borne spores, eg dry rot in wood products. If the moisture source becomes depleted the fungus may locate and transport moisture from other sources. (See *Decay* and *Mould growth.*)

Leaching Leaching is the depletion of soluble products from within a material caused by the migration of water through a material. The leached material may be transported to the surface that, on evaporation of the water, is deposited as salts. (See *Efflorescence.)*

Loss of insulation Loss of insulation will occur when the air pockets within any material fill with water. This problem is usually associated with insulating materials that have interconnecting air pockets.

Loss of strength/ disintegration Loss of strength/disintegration occurs through the softening/swelling of a material or through the loss of bond between materials in a composite in the presence of water, eg plasterboard, chipboard, woodwool etc.

Mould growth Mould growth is a superficial growth of fungus, of various colours, which commonly develops on the surface of most materials that become permanently or intermittently damp or wet, ie under conditions of high relative humidity (>70 per cent). These fungi need moisture and light. (See *Decay*, *Fungal infestation/growth* and *Staining*.)

Peeling Peeling results from the breakdown of the adhesion between a thin covering and substrate at their interfaces, eg paints, wallpaper etc.

Staining Staining is a discoloration; usually the result of mould growth or, following evaporation, of deposits transported by water to the material surface, eg rust stains. (See *Mould growth*.)

Swelling/shrinkage Swelling or shrinkage is usually the result of change, increasing and decreasing respectively, in the material moisture content. These dimensional changes may lead to warping, curling and delamination if the material is restrained or the moisture change occurs with too high a differential through the material thickness.

In addition to the above defects, control of moisture also plays an important role in the following situations:

- wood-boring beetle infestations are more likely and more damaging in damp/wet rather than dry conditions
- damp conditions both internally and externally may encourage other forms of growth, eg algae, moss.

2.2 SOURCES OF DAMPNESS

An understanding of the likely sources of damp is essential to any measurement of moisture in building elements. The determination of the source is an essential part of any investigation into moisture-related defects. Chemical analysis of water may provide important clues to the source and help highlight anomalies in moisture measurements. Alternatively adding dyes or other indicators to potential sources may help to resolve the source in some difficult situations. In an article, *Framework for defect diagnosis* (1987), Smith highlights 14 sources of dampness. The sources are discussed below.

Condensation Condensation occurs when moist air is cooled below the dew point, ie the temperature at which air becomes saturated. Condensation is very common and will usually form on cold spots within a room, eg windows, or within the construction thickness, eg roof space. Condensation is generally unsightly, leaves watermarks on evaporation and gives rise to staining, mould growth, peeling etc. When it forms within the construction thickness, it can result in corrosion of metal connectors/fasteners, timber defects and loss of insulation.

Rain penetration Rain penetration may be the result of the breakdown of a waterproofing membrane, seal, sealant or damage to the external skin of a building eg loss of tiles to roof or wall. Driving rain may penetrate a building at any location where inadequate overlap or overhang is provided eg roof or wall tiles, lead flashing and doorsills. Rainwater can also penetrate the building by capillary action particularly where deposits, eg leaves, have built up to bridge a damp proof course (dpc), blocked drains, rain splash to above dpc.

Built-in water Built-in water, or residual construction moisture, is usually associated with the wet trades, eg plasterwork, concrete, screeds, and occurs when inadequate drying times have been allowed prior to the application of an impervious coating, eg flooring. This may give rise to defects, eg blistering, peeling and mould growth. Other cases of built-in water are associated with poor storage conditions and/or inadequate protection during construction allowing the building material to become damp/saturated then not allowing adequate drying times.

Pipe leakage Pipe leakage may be the result of a burst pipe due to freezing, corrosion, joint failure, accidental damage or being overloaded.

Spillage Spillage may be the result of normal usage or design deficiencies, eg accidental while carrying a water container, carelessness/accidental while washing, pipe blockage or inadequate provision for overflow to a washbasin/bath.

Hygroscopic salts Hygroscopic salts take or give up moisture to be in equilibrium with their surroundings, eg the air, and this provides a potential source of moisture. These salts will affect test measurements made on electrical meters.

- Nitrates from decaying vegetation absorbed into the ground are significant hygroscopic salts associated with rising damp.
- Calcium chloride, sometimes once added as an admixture to concrete, is hygroscopic and will act as an electrolyte in the formation of a corrosion cell.
- Sea sand will include hygroscopic salts eg chlorides of calcium (see above), sodium, magnesium and potassium.
- Chimney damp occurs when inadequate ventilation is provided to a disused chimney that has been capped. The deposits within the chimney are hygroscopic and the chimney will be a cold spot with little in the way of a vapour barrier.
- The deterioration of magnesium oxychloride floors includes a hygroscopic build-up of moisture.
- Industrial contaminants may be present in building materials, eg salt-contaminated sand or timber.
- Animal contamination includes hygroscopic salts and also causes staining.

Rising damp Rising damp occurs near ground level and is the result of groundwater rising through capillary action in porous building materials eg brickwork, plaster. Rising damp occurs in the ground floor or walls where there is no dpc (usually pre-20th century) or damp proof membrane (dpm). It also occurs where an existing dpc or dpm has broken down, failed due to ground movement or has been bridged. Finishes will be adversely affected, eg peeling or disintegration of plaster. The salts carried by rising damp will be hygroscopic, can inhibit mould growth (important for diagnosis as it enables condensation and rising damp to be distinguished) and will affect test measurements made on electrical meters.

Seepage Seepage may occur from/in water retaining structures, eg basements, swimming pools, when the waterproofing system breaks down or fails due to movement. Test measurements may be affected.

Flooding Flooding is generally a location dependent hazard that is usually well documented. The floodwater may be from a natural source or be the result of a burst pipe, for example. It can also occur in buildings where, for example, the dpm acts as a dam to the natural flow of water. Damage will principally be confined to basements and ground floors.

2.3 MECHANISMS OF MOISTURE MOVEMENT

Moisture-related problems are often evidenced by a defect that occurs at a point remote from the source of the damp. To identify/correct the defect it is necessary to understand the mechanisms for the transmission of moisture within a building. In a temperate climate moisture is transferred from the warm, generally moist side of a building element to the cooler, generally drier side by air movement (ventilation/convection) and vapour pressure diffusion (see below). Differences in air pressure (due to wind and natural ventilation) and density (convection – hot air rises) cause air movements and are temperature dependent. Air movement has been shown to be the most significant means of moisture transfer accounting for up to 80–90 per cent of moisture vapour movement.

Other mechanisms of moisture movement include capillary action, gravity, osmosis and wind pressure, as described below.

Capillary action Due to surface tension forces, liquid water will be drawn into fine interconnecting voids, capillaries, against gravity in many building materials. Moisture is pulled in all directions by capillary action from a moisture source and is termed "penetrating damp". Rising damp (see Section 2.2 *Sources of dampness*) is the term used in the building industry when the source of moisture is groundwater.

Gravity Gravity ensures that water will normally run downwards. Small quantities of water will, however, form droplets and adhere to surfaces due to surface tension effects. To prevent significant build up of water, eg ponding on a flat roof, a fall of 1 in 80 is considered the minimum required slope. Due to construction inaccuracies and short- and long-term deflections design, minimum falls of 1 in 40 are generally recommended.

Excess water will eventually form ponds or reservoirs at barriers/restrictions to flow and, with increasing depth, build up water pressure, eg groundwater pressure on foundations/ basements. Water under pressure may find weaknesses in water proofing systems and leak into the building fabric.

Osmosis Osmosis is the diffusion of a solvent, in this case water, through a semi-permeable membrane into a more concentrated solution, tending to equalise the concentrations. This has resulted in blistering to floor coverings on concrete ground floor slabs when, for example, incomplete curing of an epoxy adhesive has formed a semi-permeable membrane at the concrete surface. The blistering usually appears as small bubbles, randomly spaced and containing a salt solution. A chemical analysis of the solution will aid identification of the osmotic process.

Vapour pressure diffusion Water is naturally held as vapour in the air. The vapour pressure of air, at a given temperature and relative humidity, is quantifiable from psychrometric charts/equations. When air on the inside of a building has a different vapour pressure to that on the outside, moisture is driven through the building fabric by that differential pressure to reach equilibrium.

Wind pressure Rain penetration can often be attributed to wind. Rain is driven against the outside windward faces and may be forced, even against gravity, through overlaps/overhung openings, defective seals or joints. Buildings also create barriers to wind flow setting up pressure differentials, including suction that can result in rain penetrating the building fabric.

2.4 ELEMENTS AND ENVIRONMENTS

The excess or absence of moisture in the environment is likely to be a key factor in the performance of building elements. When investigating failures or potential defects, it is important to recognise the need to conduct a comprehensive visual survey in which the element type and environment conditions will normally be considered together. External elements will need to be considered in relation to the prevailing wind direction, orientation (exposure to sunlight) and location (climatic, urban etc, height, and/or coastal). Internal elements also have a wide a range of potential exposure conditions, ie building use (domestic, office, hospital etc), general usage (office, kitchen, bedroom etc) and environmental control (heating, air conditioning etc).

It is important to remember that people of necessity generate heat and moisture, resulting in differential temperatures and air and vapour pressures internally and externally. These differential pressures will drive the air, containing moisture vapour, through most building materials. If, during this transfer, the temperature drops at any interface (air/material or different materials) to the level where the air becomes saturated, ie reaches the dew point, the water vapour will condense at that location. It is in this liquid form that water potentially gives rise to most problems. Adequate ventilation is the best control to prevent condensation from becoming a problem.

It should also be recognised that excess or depletion of moisture in the building environment may cause discomfort to the occupants, eg respiratory ailments, and may be detrimental to the operations carried out in the building. This may lead to complaints under the Environmental Protection Act as producing conditions unfit for habitation.

2.5 MATERIALS AND INTERFACES

Most building materials contain moisture. Problems arise when the quantity, the moisture content, falls or rises outside prescribed limits. Knowledge of actual moisture content can therefore be seen as desirable with the provision that acceptable moisture content limits for different building materials vary considerably, for example at 6 per cent wood is "very dry" and concrete is "saturated". Rate of change and differential moisture content within a material can be as important a consideration as actual moisture content. (See Section 2.1 *Effects of moisture.*)

Hygroscopic materials such as wood have the characteristic that they can either take or give up moisture to the air to achieve a balance with the prevailing relative humidity. This is termed "equilibrium moisture content". Hence during apparently normal building operational conditions these materials may be adversely affected by moisture from water vapour transmitted by air movement. (See Section 2.3 *Mechanisms of moisture movement.*)

As with other building defects, moisture-related problems tend to occur at or in the connection at interfaces, either at the surface (air/material) or between materials. The reason is that condensation generally forms at interfaces and moisture tends to migrate and collect at interfaces between permeable and impermeable materials.

3 Testing for moisture

3.1 OBJECTIVES FOR TESTING

There are two main objectives for testing: to determine firstly the severity or degree of moisture present and secondly the source of the moisture.

Severity/degree of moisture

Absolute moisture test data provides essential information that may help to identify how the moisture is held, ie to distinguish the "free" moisture (see Section 3.3 *Moisture measurements*). This may aid evaluation of the probable current and future risks.

Comparative data can provide an invaluable diagnostic tool to identify particular defects. For example, moisture gradients provide a pointer to the source to distinguish between the moisture problems caused by rising damp and condensation, ie to differentiate between defect types. It is important that all potential sources of the moisture be considered in any investigation and confirmed or eliminated (see Section 2.2 *Sources of dampness*).

Source of moisture

Source of moisture testing usually implies the existence of a defect, eg a leak. The first sign of water penetration within a building may not be of much assistance in identifying the source or entry point; the path water takes within a building is often unpredictable and may be tortuous. The test technique may be based on detecting the presence of unwanted moisture or in identifying the location of defects in the waterproofing system. The latter technique can be also used for quality control to enable certification of the absence of such defects.

3.2 REASONS FOR TESTING

Prevention is better than cure, hence quality control testing to demonstrate the minimisation of risk and long-term monitoring to give early warning of potential moisture-related defects should be encouraged. In order for this approach to be effective, the person specifying the tests needs to have a fundamental understanding of the parameters involved. The measurements need to be analysed and the data well presented in order to derive the full benefit of testing.

The aim should be to provide the test specifications for quality control during construction, operation and maintenance and, if required, for long-term monitoring. If properly implemented this should minimise moisture-related defects in building elements. Failure to do so may render the designer/contractor liable for defects.

The three reasons for testing are to gain information on moisture presence in building elements, ie quality control, long-term monitoring and defect investigation. They are considered below. Currently most testing is conducted to investigate problems.

Quality control

With the increasing emphasis on "doing things right first time", practical methods which can be used to establish/confirm that construction work is satisfactory are increasingly necessary, ie quality control testing during construction. Building owners may also require certification of compliance as part of quality assurance. Examples applications are given in Table 3.1.

Table 3.1 *Example applications of testing for quality control*

Example application	Test method
Timber has been provided and installed in a specified moisture condition	electrical resistance or capacitance
Concrete/screed floors have dried to a moisture condition to allow impervious covering to be laid safely	humidity sensors and electrical resistance or capacitance for pre test or to check stability
Plaster/walls have dried to a moisture condition to allow finishes to be safely applied	electrical resistance or capacitance, calcium carbide moisture meter
Demonstrate that condensation will not have formed on surfaces, eg steelwork before painting	(temperature) and humidity sensors
Flat roofs have been constructed without defects	electrical capacitance, earth leakage, thermographic inspection
Joints/gaskets/seals and sealants have been correctly installed	(air pressure tests) or ultrasonics

Long-term monitoring

Long-term monitoring can be achieved by two alternative strategies:

- repeat manual measurements over the required time period
- automated measurements (see Section 6 *Automated long-term monitoring*).

Long-term monitoring has been used on many buildings, including historic/special-interest buildings, where potential damage, due to moisture to the building fabric or the building contents, is unacceptable. Example applications are given in Table 3.2.

Table 3.2 *Example applications of testing for long-term monitoring*

Example application	Sensor	Auto	Repeat
A periodic check on the moisture condition building fabric	electrical resistance or capacitance meter	no	yes
To determine the efficacy of accelerated drying after fire or flood	embed humidity sensors within the building fabric	yes	yes
To check the efficacy of remedial works or materials	(install resistance probes in the material)	yes	yes
To check the integrity of roofs	(leak detection tapes)	yes	no
To detect failure or leaks in plumbing systems	(install flow meter and monitor usage patterns)	yes	no

Repeat periodic measurements have advantages:

- relative low initial and maintenance cost of comparable equipment
- potential for visual inspection while taking measurements

and pitfalls:

- lack of potentially critical data between visits
- the staff time and costs of access
- disruption of occupancy to obtain measurements
- possible health and safety risks to staff in carrying out measurements.

Any sensor that gives an electronic signal can be linked to an automated monitoring system to provide real-time early warning of potential defects and historic data. Building owners/operators will only derive the main benefits from the system if the installation has been well designed, is properly maintained and the data continuously processed. Obviously, circumstances exist where the use of automatic monitoring systems would not be cost-effective. Firstly, the installed cost of the sensors must be less than the cost of resolving the problem in some other way. Secondly, the data gathered must be effectively used to control the problem.

Defect investigation Virtually all moisture-related problems are characterised by a visual appearance that can be used to aid interpretation. Prior to carrying out any defect investigation testing, an essential prerequisite is that the historical information needs to be compiled in particular, design, construction, maintenance and repair details. An understanding of the conditions during construction also plays a major role in the assessment of problems during installation and occupancy, even after considerable time if chemical damage is involved. This includes internal and external environmental conditions and previous and current use of the land and buildings. An equally important prerequisite to testing is a site visit to confirm the background information and objectively investigate and record the visual evidence. Information should be sought for the potential sources of dampness, as these can be varied. It may simply be the residual construction moisture, the residual water from subsequent operation or from accidental spillage/leaks.

When investigating a defect, testing for moisture provides essential clues to the nature and cause and hence the required repair strategy. Example applications are given in Table 3.3.

Table 3.3 *Example applications of testing for defect investigation*

Construction	Test method	Commentary
Roof membranes (flat) exposed	electrical earth leakage	tracing leak source
	electrical capacitance	entrapped moisture/tracing leak source
	(localised flood/ dye testing)	pointers to defective details of outlets and downpipes
	thermographic inspection	can pick up deficiencies in insulation; the highest loss of insulation can be coincident with the leak source
Flat roofs, insulated and ballasted	possibly electrical capacitance/electrical earth leakage	depending upon nature of membrane: a degree of preparation is required for this test but results can often be obtained without lifting all the insulation and ballast.
Other buried membranes	(flood/dye testing in the first instance)	if this is successful, removal of finishes to expose membrane
	electrical earth leakage/ electrical capacitance	tracing leak source/entrapped moisture
Metal roofs, insulated, pitched	thermographic inspection	can often show loss of insulating characteristics due to water paths within the roof
	(localised visual inspections)	as verification where there is translucent sheeting in place, the junctions between plastic and metal can often be the cause of leakage
	(spray/dye testing)	moisture source confirmed likely/ruled out by test results
Brick, block or concrete walls	electrical capacitance or resistance/oven drying/ calcium carbide moisture meter/humidity sensors	moisture content analysis to determine severity, pattern and potential source of dampness
	thermographic inspection	can help to determine the extent of the area of dampness/loss of insulation within the construction
Cavity walls	(borescope inspection)	can often determine if the tray is correctly installed and whether it is functioning correctly
Concrete/screed floors	(visual inspection/ breakout investigation)	to determine failure mechanism/construction detail, expansion and contraction joints (restraint), edge details
	electrical capacitance or resistance/oven drying/ calcium carbide moisture meter/humidity sensors	when investigating flooring failures, moisture content analysis to determine severity, pattern and potential source of dampness
Timber cladding/ flooring	(visual inspection)	to determine failure mechanism, evidence of fungal or insect infestation
	electrical capacitance or resistance/humidity sensors	moisture content analysis to determine severity, pattern and potential source of dampness

See Section 2.2 *Sources of dampness* and Section 4.1 *Diagnostic strategy/procedures*.

3.3 MOISTURE MEASUREMENTS

Before selecting test method(s) for moisture measurements it is important to decide:

- what type of data is required for the particular situation and test objective
- the factors which must be taken into account to ensure the consistency and accuracy of the data
- other properties that need to be tested to interpret the results correctly, eg temperature, density
- whether the use of more than one test method would be appropriate to reach the required results.

Factors affecting measurements

There are many factors that can affect moisture measurements and the interpretation of the results:

- the majority of techniques are "indirect" in that they measure some parameter or property of the material which changes with moisture content, for example: electrical or heat conductivity, changes in dielectric constant or absorption of energy from an electromagnetic field. However, any other material(s) present which cause similar changes on the measurement to that of water, eg salts, will affect the reading obtained and thus the accuracy of the moisture measurement
- the interpretation of the results is further complicated by the fact that water can be present in many forms (see Figure 3.1). Most measuring techniques do not distinguish between or measure all of these different forms; hence different equipment may give very different results for the same sample
- the effective depth at which the measurement is made which may be dependent on the test technique

Figure 3.1 *Schematic showing how water may be present*

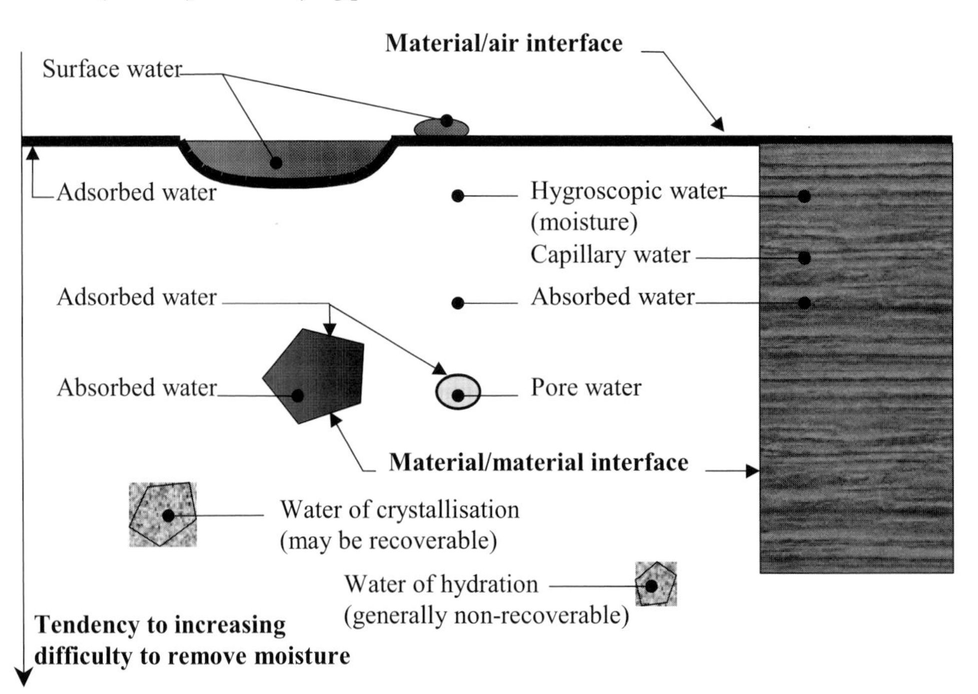

- many of the techniques are affected by the nature and density of the material under consideration: if the element/material being measured is not of a consistent construction, the results will not be comparable, eg knots in wood, roofs of different construction
- the effect, if any, of surface moisture on the test measurement
- temperature can also affect measurements taken, eg humidity, resistivity
- the material or medium being measured may not have reached equilibrium.

In view of the above, it is clearly essential when considering moisture measurement and the degree of accuracy required to know precisely what each instrument measures, and what other factors need to be taken into account. This is especially important if the readings are to be compared with those from another technique, particularly in understanding the limitations when calibrating one instrument by the results of another, eg calibrating an electrical capacitance or resistance meter against the results of the calcium carbide moisture meter method.

Type of measurement

The moisture condition of building materials can be expressed either in absolute or comparative terms. Absolute values of moisture content provide measurements that can be traced to national standards. Comparative measures allow identification of the moisture conditions within areas of the same material, either in similar numeric units to an absolute measure (but without traceability) or by an indirect measurement.

Presentation of results

Whether the measurements are in absolute or comparative terms, the data can be presented in tabular form or plotted to provide moisture profiles, eg through the construction thickness or by mapping the data from a given area and adding moisture contours. Selecting the appropriate graphical presentation will considerably aid interpretation of the data and may significantly improve communication of its meaning.

Moisture content

Moisture content (MC), determined by oven drying in a laboratory, is the only absolute measure of the quantity of water in a material (see Section 5.9 *Oven drying method*). In addition to the factors above, the following problems also need to be considered when measuring moisture content and interpreting the results by this method:

- particular care needs to be taken when taking samples by drilling to ensure that moisture is not driven off by heat generated in the drilling and the sample is not contaminated by material from other depths
- samples, particularly small samples, may not be representative of the bulk of the material; as a guide, eg for concrete, the sample volume collected should be at least 200 times the volume of the largest particle. This sample may need to be processed, ie divided into smaller and now representative sub-samples for individual tests
- the drying process may drive off substances other than water.

Other test techniques may provide direct or indirect moisture content measurements that are calibrated for specific materials and conditions. For further details refer to the specific test methods in Section 5 or to the manufacturer's literature. **It is very important to understand the limitations of such calibrations.**

The moisture content of building materials is usually expressed as a percentage of either the final dry or original wet weight of the material. It is occasionally expressed in terms of a percentage of the respective volumes, again of either wet or dry volume. Care must be taken not to mix terms or confuse relative (a percentage from an arbitrary meter scale) with absolute readings. Additionally hygroscopic moisture content (HMC) and/or chemical tests can be carried out to determine if the sample contains agents eg salts that might influence interpretation of the data.

Common building materials generally have low hygroscopicity and are tested for HMC reaction in 75 per cent relative humidity that should expose hygroscopic contaminants, ie at a rh corresponding to some of the worst conditions likely to prevail inside property in the UK. BRE Digest 245 refers to a 5 per cent wet weight moisture content in masonry materials as representing the threshold of problematic conditions; the 5 per cent threshold being a mix of capillary moisture and hygroscopic moisture. However certain materials may possess a HMC of up to 5 per cent without hygroscopic salt contamination, which emphasises the need when carrying out an investigation to test for both MC and HMC. It is also important to appreciate that in typical masonry materials the maximum HMC expected at 75 per cent rh would be less than 2 per cent, (Johansson and Persson, 1998) leaving the suggested problematic capillary moisture content at up to 3 per cent. In the evaluation of dampness problems, BRE Digest 245 is very helpful in suggesting procedures to distinguish between various causes of dampness in order that correct diagnosis and appropriate remedial action may be taken. In masonry walls it is suggested that the mortar is a significant material for testing.

There is considerable variation in the quantity of water that specific building materials can retain before becoming saturated. Concrete, for example, will usually have moisture contents in the range up to 6 per cent, limited by the capacity of the pores: wood on the other hand will generally have moisture contents of 6–40 per cent, both expressed in terms of dry weight.

Testing for moisture content in absolute terms requires samples to be taken that result in minor damage to the element. The results can be used to verify moisture meter readings thus minimising the need for taking samples when carrying out major investigations. Care must be taken to ensure that the results from these different methods are directly comparable (see previous sub-section on *Factors affecting measurements* and Section 5 *Test methods*).

Comparative measurements are expressed either in similar numeric units to an absolute measure (but without traceability) or in units appropriate to the parameter being measured, (eg kohm cm for electrical resistivity readings).

Moisture content may also be expressed in the following terms.

Wood moisture equivalent

Wood is a hygroscopic material, ie will take up or release moisture to remain in equilibrium with its surroundings. As a generalisation, if conditions give rise to wood moisture contents above 20 per cent problems are likely to occur either in the wood or the surrounding materials. This has given rise to the concept Wood Moisture Equivalent (WME – intellectual property of Protimeter plc) by which the moisture level of any building material is expressed in terms of the moisture content of wood that is in close contact and in moisture equilibrium with the material. A major problem is "calibration" for the material under test and the main disadvantage is that the data quoted may be confused with absolute measurement values (see Section 5.*4 Electrical resistance meter*).

Relative humidity

The moisture content of a building material may also be measured by recording the relative humidity within an enclosed air space created either within or on the material surface: it may also be referred to as equivalent relative humidity. The measurement of relative humidity is well documented and there are a number of commercially available sensors and meters. The main drawbacks are the time required to reach equilibrium, which may be several days, and the need for frequent re-calibration (see Section 5.5 *Humidity sensors*). The advantage is that it represents a wider area and depth of material than is usually tested by other techniques, ie more representative since it reflects the moisture condition of the bulk of the material surrounding the air space.

3.4 STANDARDS AND SPECIFICATIONS

There are few UK or European standards or specifications that specifically cover the test methods for moisture measurement in building materials. However some companies work to BS EN ISO 9002: 1994 *Quality systems. Model for quality assurance in production, installation and servicing* that should ensure consistency of approach when taking moisture measurements.

Standards and specifications that recommend the use of particular test methods have been identified in the relevant sub-sections of Section 5 *Test methods* and also Section 7 *References and bibliography*.

3.5 HEALTH AND SAFETY

Construction (Design and Management) Regulations 1994 (CDM Regulations) require the production of documents for health and safety plans and a health and safety file for the majority of construction projects. These documents should demonstrate that reasonable care has been taken to manage health and safety issues and that these have been taken into account. They should comprise a method statement and risk assessment for the work to be carried out. These will make reference to personal protective equipment (HMSO, 1992), manual handling operations (HMSO, 1992) and, if handling or working with potentially hazardous substances, to Control of Substances Hazardous to Health (COSHH) data sheets (HMSO, 1994). An important part of the management of risk is that the parties should agree the method statement and risk assessment. They should also demonstrate that reasonable care has been taken to ensure that all personnel involved with the work have been fully briefed and understand the documents.

The Health and Safety at Work etc Act 1974 requires all organisations (and individuals) to safeguard the health and safety of those who carry the work and any person it may affect. General health and safety issues regarding working on building sites should be applied even when the building is occupied, for example:

- access, particularly onto roofs
- working at heights
- protection to operator and others against falling objects
- preventing trip hazards
- blocking fire escape routes.

Manufacturers instructions should always be read and observed and particularly care should be taken when handling unknown/hazardous materials, eg when taking samples, re-instating damage.

Section 5 *Test methods* give particular health and safety issues for each test method.

3.6 TRAINING/CERTIFICATION

Although some of the equipment is comparatively easy to handle and read, in order to obtain correct and meaningful results it is essential that an experienced person undertakes or supervises the testing and data interpretation. This technical competence should be verified before commissioning testing. Short-listed companies should give UKAS or equivalent accreditation or, as a minimum qualification, provide a demonstrable track record for the testing work to be undertaken. The BWPDA provides a certification scheme for operatives.

3.7 A FINAL WORD

The selection of one or more test methods should aim to ensure that the test results provide the required information.

Quality control: to provide meaningful data to demonstrate the minimisation of potential problems and the absence of built-in defects, ie "right first time".

Long-term monitoring: to give early warning to eliminate/minimise damage by timely maintenance/repairs.

Defect investigation: to assist identification of the defect, likely source of the moisture and mechanism of moisture movement, ie "put right first time".

Figure 3.2 illustrates the interrelationships between the parameters and the problems which face anyone dealing with moisture-related effects on building elements.

If there is a temptation to jump directly to another section, pause a moment to reflect. Reaching conclusions based on incomplete analysis of moisture-related effects have resulted in millions of pounds of unnecessary expenditure on buildings to "cure" moisture-related effects, for example:

- the chemical injection of ground floor walls showing any form of dampness
- the wholesale removal of "sound" materials showing indications of/in close proximity to dry rot.

Figure 3.2 initially aims to illustrate four main features in the process of selecting the best test method(s) for the measurement of moisture in building elements:

- the main parameters that need to be considered
- there is no defined starting point
- there is no single or correct route
- it is all too easy to make a test selection without consideration of all the parameters and then, using just this incomplete data, jump to the wrong conclusion!

The figure has the orientation with the recommended starting point at the top which should be considered during the design and construction phases, ie to provide the test specifications for quality control during construction, operation and maintenance and, if required, for long-term monitoring to provide early warning of potential problems. If this route is followed, the figure emphasises:

- the strong link between element with environment and material with interface
- that the test method selection for quality control and long-term monitoring has links to, but is remote from, defect identification.

Currently the most common route for test selection starts from the bottom of Figure 3.2, with an observed defect. Starting the selection route from this perspective emphasises:

- that researching background details and visual examination are the first and, arguably, the most important requirements in any defect investigation
- the close link between a moisture-related defect, the source of moisture and the need to understand the moisture transfer mechanism.

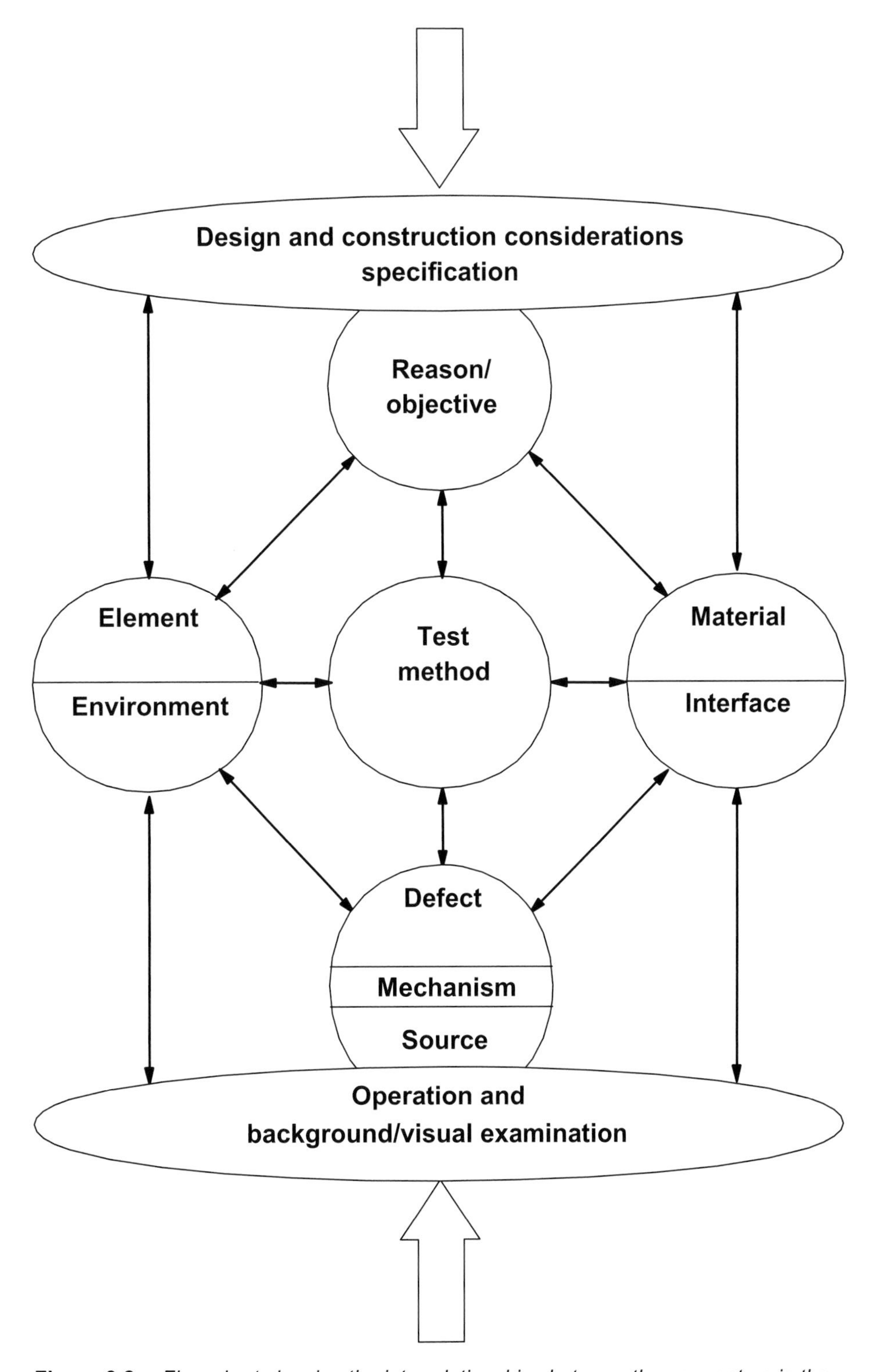

Figure 3.2 *Flow chart showing the interrelationships between the parameters in the process of selecting test methods for moisture measurement*

It is not inappropriate that the figure looks like an hourglass that literally turns on completion of the building and runs during the building lifetime to essential maintenance. Following effective maintenance/repair work, the timer is reset.

4 Selecting the appropriate test method

4.1 DIAGNOSTIC STRATEGY/PROCEDURES

An analytical approach to defect diagnosis

Water is often described as the "elixir of life" but, in the wrong place, it can be an insidious material, moving upwards, downwards and sideways in buildings. It can also travel long distances before manifesting internally as dampness.

When a defect occurs, it is vitally important that it is resolved in the most efficient and cost-effective way and the reasons ascertained for its cause. If not, the investigative process can be prolonged and repairs ineffective.

Many materials are involved in the building process and it is often the marrying of two or more of these into any particular element that is the root of the problem.

It is vital that a staged approach is taken in the analytical process, each stage possibly having a number of variables. The results from each stage will determine the considerations and variables to be taken forward to the next. A four-stage approach to problem solving is recommended (Concrete Bridge Development Group, Technical Guide No 2, *Testing and monitoring the durability of concrete structures,* to be published).

1. Definition.
2. Identification.
3. Analysis.
4. Confirmation.

A typical approach to the analytical process might be as described below.

Stage 1: Definition

Try to answer the questions: "what, where, when and how big is the problem?"

- The nature of the dampness needs to be clearly defined – is it continuously wet with water present, intermittently wet, or just damp?
- When does the dampness appear, is it weather- or use-related?
- How long has the dampness been apparent?
- What repairs, alterations or changes in use, if any, have taken place? Are they considered to be relevant? Did they occur prior to the identification of the problem?
- If it appears to be weather-related: how long after rain starts, is it associated with wind-blown rain, how long does it continue once the rain stops? What colour is the water? Does it change according to the quantity? What volumes of water are collected during rainfall?
- If it appears to be use-related: what is considered to be the potential source of the dampness? Is there any evidence to confirm the source and/or eliminate any other source?

Stage 2: Identification *Try to identify the distinctive characteristics of, and any changes that may have contributed to, the problem.*

- If available, obtain the building records, not just design and construction details but the Health and Safety file, operation and maintenance and repair records. Details of any alterations or additions and the previous site history may also be relevant. For older buildings this may not be possible; building records have either been lost or are archived beyond the reach of the investigator.
- Try to identify and locate all potential sources and trace the likely routes and mechanism of the moisture transfer to the problem areas:
  - is there a plant room wall or services entry directly above or close that might provide the moisture entry point?
  - is it close to water mains, drains, rainwater gutters or down-pipes?
  - is it near to gaskets or seals at openings or construction joints?
- Obtain as much superficial information as possible, for example:
  - what type of roof decking has been used?
  - is the roof insulated?
  - are there cavities within the brickwork and do they require trays to keep them watertight?
  - is the curtain walling free draining?
  - what type of dpc or dpm has been provided?
- What was the floor construction?

Stage 3: Analysis *Get to grips with the problem by a process of elimination.*

- From the information obtained in Stages 1 and 2, set up likely scenarios and identify the best way to confirm or eliminate them from the investigation. For this process, a working knowledge of construction techniques and an understanding of moisture in building elements are vital; experience is essential.
- Create an inspection plan bearing in mind not to make the investigative process unnecessarily complicated, some problems can be resolved by visual inspection. For this reason it is usually best to start the investigation by recording the visual evidence.
- How does the dampness relate to items above or beside it and could these be detailed or installed incorrectly? During the investigation keep an open mind and if the information gained does not correspond to that identified in Stages 1 and 2, revise the plan accordingly.
- Whilst visual inspection can take the investigation so far, moisture mapping or profiling using non-destructive comparative test methods can provide indicators of moisture paths that are invisible to the naked eye. It is usually best to trace the moisture path back internally as far as possible. In this way, investigations or tests based on destructive techniques may be directed at items closest in relation to the internal points of highest moisture density to obtain more quantitative information, for example chemical tests can be used to discriminate between mains water, rainwater and sewage.
- Start to eliminate scenarios one by one if possible, although two or three often function together. Depending on results, these can be eliminated as a group or segregated down for further investigation.
- Beware of jumping to conclusions before eliminating other likely scenarios.

Stage 4: Confirmation

Make sure the diagnosis is correct.

- If it is necessary to prove a scenario by carrying out flood, spray or dye testing it is best to isolate construction elements from the lowest point up, in order to minimise additional potential damage. Also, should this testing prove negative, it should be designed not to hinder/obscure further investigations.
- It may be necessary to carry out additional testing to monitor the effectiveness of any remedial treatment and hence confirm the diagnosis.

The four-stage approach to problem solving is equally applicable to quality control and long-term monitoring.

4.2 EXAMPLE APPLICATIONS

Table 4.1 illustrates that quality control and long-term monitoring testing should be considered together in the overall test strategy. It also highlights the need for a strong link to the design and construction detail.

Similarly, Table 4.2 illustrates the link between defect investigation and long-term monitoring. Selecting the test method should not be considered in isolation; it should form part of an overall strategy.

Table 4.1 *Example application for quality control and long-term monitoring*

Scenario	How to use guide	Stage
Quality control specification required for timber flooring.	**Table 3.1** gives example applications of testing for quality control. **Tables 3.1** and **4.4** identify the appropriate test methods.	1
A parquet flooring comprising 600 mm × 45 mm × 9 mm thick strips of narra rose-wood is to be laid in a new but unoccupied community building.	**Section 2** explains the relationship of materials with their interfaces and elements with environment. Wood is hygroscopic and will change moisture content to be in equilibrium with its environment – consider how to control and/or accommodate shrinkage/swelling. **Tables 4.4** and **4.5** identify test methods for wood moisture condition to be checked. Front runner electrical capacitance meter as non-destructive, resistance meter leaves pin-holes – more detailed descriptions in **Section 5**.	2
The parquet flooring is to be laid on sand cement screed over a concrete sub floor.	**Table 4.4** and **4.5** identifies test methods for screed moisture condition to be checked. Front runner humidity sensors – more detailed description in **Section 5**.	
The ground-bearing slab will be installed with polythene dpm. To speed construction the slab is to be overlaid with applied dpm before screeding and room heating run to design levels.	**Section 2** details likely problems with wood and screeds, likely sources of moisture and mechanism for moisture movement in ground floor slabs. The dpm will prevent moisture rising from the concrete slab. The room heating may aid the screed to reach the required moisture condition in the required time scale. However the primary consideration needs to be the provision of strong ventilation to remove the humid air from the building. If this is not achieved the moisture will be retained in other parts of the building fabric with attendant high risks of condensation and the timber swelling causing disruption to the flooring. Consider installing temperature and humidity sensors to control heating and ventilation during construction.	3
Long-term monitoring.	**Tables 3.2** and **4.4** identify test methods for long-term monitoring applications. Humidity sensors/(resistivity probes) front runners for this application.	4
Unoccupied community building.	Consider continued use of temperature and humidity sensors to control and monitor heating and ventilation during building life coupled with humidity sensors embedded in the building fabric. There is the need for co-ordination at the design stage when considering long-term monitoring to minimise disruption of the building fabric to install sensors and route cables.	

Table 4.2 *Example application for defect investigation and long-term monitoring*

Scenario	How to use guide	Stage
The occupants of a five-storey office building constructed 15 years ago are reporting damp patches appearing on ceiling panels in fifth floor rooms. Roof leaks are suspected.	**Section 2** gives descriptions of effects of moisture (defects) and **Section 4** the strategy/ procedures for defect investigation.	1
The building has a flat roof with central plant and lift motor room. A "warm" roof construction has been used with mastic asphalt on separating membrane to polystyrene insulation, vapour barrier and vented mat on screeded concrete deck. A built-up felt roof construction has been used for the plant/ lift room.	**Section 3** and **4** identify the need to seek construction details that may need to be confirmed during site investigation. Consider what equipment is required for this part of the investigation. No details of any alterations, maintenance and repairs provided. **Table 4.5** identifies electrical capacitance and earth leakage as the likely front-runners to locate any leak in the roof, or possibly use of thermographic inspection or radar. **Table 4.4** identifies electrical capacitance or resistance meters for measurement of material moisture content. Query the need for calcium carbide moisture meter/oven drying to determine absolute moisture levels at this stage. **Section 5** provides further guidance.	2
Damp patches at ceiling level have become evident in several rooms during the winter season.	**Section 3** and **4** identify the need to seek background history of the defect to be investigated during a site visit.	
A visual inspection reveals that damage to ceilings is restricted to the top floor near the lift entrance. The damage to the ceiling correlates strongly with extensive water penetration of the built-up felt roof to the plant room. Additional direct water penetration has occurred through rotted doors and thresholds to the plant room. The main roof waterproofing had not been carried through to plant room floor. The asphalt to the main roof appeared to be sound, minimal surface cracking.	The main roof is probably sound. Suspend further investigation of the main roof. Carry out remedial work to plant room roof and doors to eliminate the likely source of moisture. Ventilate affected ceiling void to minimise the risk of further defects by sealing within the structure the moisture that has penetrated the building fabric. Recommend better maintenance strategy and compiling building maintenance records (see BS 8210: 1986 *Guide to building maintenance management*).	3
Long-term monitoring.	Consider need for **long-term monitoring** to evaluate efficacy of the ventilation programme. **Section 3** and **Table 4.4** identifies test methods, humidity sensors are front-runners for this application.	4

4.3 CRITERIA FOR TEST SELECTION

Table 4.3 lists the parameters that need to be considered when selecting the appropriate test method(s), together with reference to the appropriate section or table in which guidance can be found.

Table 4.3 *Parameters to be considered when selecting test methods*

Parameter for consideration	Refer to:
Reason/objective for testing	Sections 3.1 and 3.2 and Table 4.4
Non-destructive or some damage acceptable (in-situ or laboratory)	Table 4.4
Absolute or comparative data	Section 3.3 and Table 4.4
Element and environment	Section 2.4
Material and interfaces	Section 2.5
Defect type(s)	Section 2.1 and Table 4.6
Source(s) of moisture	Section 2.2 and Table 4.7
Mechanism(s) of moisture movement	Section 2.3
Testing for moisture near the surface, at various depths within material thickness or in substrate material under a finish	Section 3.3
Number and position of test locations	Section 5
Test for moisture only or for other performance/ physical characteristics/chemical type present	Section 4.1
Access requirement to building element and accessibility at test locations	Section 5
Cost and turn-round time	Section 5

4.4 TEST SELECTION TABLES

The tables given below were developed with **the aim to rank test methods for problem solution**. This should aid identification of the test methods that can be used to confirm any initial diagnosis and hence identification of the solution. The ranking method employed has four levels, as follows:

3	highly applicable
2	sometimes applicable
1	only rarely applicable
(Blank)	not applicable.

The rankings will help to identify relevant test methods for a given situation, but not to differentiate between them nor which type and model to use. The tables only refer to the use of the methods that are detailed in Section 5. Reference to the authored text for these test methods will provide the next stage in the selection process. Further detailed information on the appropriate use of these and other test methods should be sought from consultants/test house specialists in moisture-related problems.

Table 4.4 *Links **objective/reason** of test and **test method***

Test method	Calcium carbide moisture	Electrical capacitance	Electrical earth leakage	Electrical resistance	Humidity sensors	Microwave moisture meter	Nuclear moisture gauge	Nuclear magnetic resonance	Oven drying method	Radar	Thermographic inspection	Ultrasounics
Place of measurement (i)	I	I	I	I	I	I	I	L	L	I	I	I
Damage caused (ii)	D	N	N	S/D	N/D	N	N	D	D	N	N	N
Measurement (iii)	A	C_A	C	C_A	C_A	C_A	A	A	A	C	C	C
Objective/reason												
Severity/degree of moisture	3	3		3	2	1	2	1	3	1		
Source of moisture	2	3	2	3	1		2		1	1	1	2
Quality control and repeat testing	3	3	3	3	2	1	2	1	3	2	1	2
Automated long-term monitoring		2		3	3							
Defect sort/cause	2	3		3	2	1	2		2	2	1	2

Notes

(i) I = in-situ, L = laboratory

(ii) N = non-destructive, S = semi-destructive (pin-holes), D = destructive

(iii) A = absolute, C = comparative, C_A = can also be absolute.

Notation The ranking method for problem solution has four levels:

3	highly applicable
2	sometimes applicable
1	only rarely applicable
(Blank)	not applicable.

Table 4.5 *Links **element**, **material** and **test method***

Element	Material	Calcium carbide moisture	Electrical capacitance	Electrical earth leakage	Electrical resistance	Humidity sensors	Microwave moisture meter	Nuclear moisture gauge	Nuclear magnetic resonance	Oven drying method	Radar	Thermographic inspection	Ultrasounics
General													
	aggregate	2					2	1	1	3	1		
	concrete	2	3		3	3	2	1	1	3	2	2	1
	mortar	3	2		3	3	2	1	2	3	1	2	
	sand	3				2	3	2	1	3	1		
	soils	3				2	3	3	1	3	2	1	
	wood		3		3	2		1	1	2	1	1	1
	stone	2	3		3	2	2	2	1	3	2	2	
Floors													
	screed	3	3		3	3	2		1	2	2	2	
	substrate under covering	1	3	1	3	3		1	1	2	2	1	
Walls/ceilings													
	brickwork/blockwork	3	3		3	2	2	1	1	3	2	2	1
	plaster	3	3		3	3	2		1	2	1	2	
	plasterboard		3		3	2			1		1	2	
	render	2	2		2	3	2		1	3	1	2	
	glazing/joints/sealant											1	2
	substrate under paint		3	1	1	1					1	1	
	substrate under tile		3	1	2	2		1	1	2	2	1	
	substrate under covering		3	1	2	1					1	1	
Roofs													
	membrane without conductive layer [(i)]		3	3	1	2	1	1			1	2	
	membrane with conductive layer [(ii)]		2		1	2	1				1	2	
	mineral wool		3		1	3	1	2		3	1	2	
	woodwool		3		1	3	1	2		2	1	1	

Notes (i) pvc single-ply, EP single-ply, bitumen felt, asphalt, liquid coating

(ii) metal-faced felt, EPDM or butyl single-ply.

Notation See Table 4.4.

Table 4.6 *Links **defect** and **test method***

Defect / **Test method**	Calcium carbide moisture	Electrical capacitance	Electrical earth leakage	Electrical resistance	Humidity sensors	Microwave moisture meter	Nuclear moisture gauge	Nuclear magnetic resonance	Oven drying method	Radar	Thermographic inspection	Ultrasonics
Blistering	2	2		2	3	2			2			
Corrosion		2		2	2					2		
Damp/wet	3	3		3	3	2	2	1	3	2	2	2
Decay		3		3	2			1	3			1
Delamination	2	2		2	2	2		1	2	2	1	
Freeze/thaw damage	2	2		2		2		1	2	2	1	
Fungal infestation/growth		2		2	2			1	2			1
Leaching	1	1		1	2				2	1		
Loss of strength/ disintegration	1	1		1	2			1	2	1		
Mould growth	1	1		1	1				2			1
Peeling	2	2		2					2			
Staining	2	2		2	2			1	2			2

Notation The ranking method for problem solution has four levels:

3 highly applicable

2 sometimes applicable

1 only rarely applicable

(Blank) not applicable.

Table 4.7 *Links **source** of moisture and **test method***

Source / **Test method**	Calcium carbide moisture	Electrical capacitance	Electrical earth leakage	Electrical resistance	Humidity sensors	Microwave moisture meter	Nuclear moisture gauge	Nuclear magnetic resonance	Oven drying method	Radar	Thermographic inspection	Ultrasounics
Animal contamination		1		1								
Built-in water	3	3		3	3	2	1	1	3	1	1	
Calcium chloride		1		1					3			
Chimney damp		2		2	2				3		1	
Condensation	1	2		2	3				1	1	2	
Deterioration of magnesium oxychloride floors	2	2		2	1	1			2			
Flooding	2	3		3	3	1	1	1	2	2	1	
Industrial contamination		1		1						1		
Pipe leakage	2	3		3	2	1	1	1	3	2	2	
Rain penetration	1	3	3	3	3				1	2	2	3
Rising damp	2	3		3	2	1	1	1	2	1	1	
Sea sand contamination		1		1					3			
Seepage	1	3	3	3	2				1	2	2	
Spillage (including washing)	2	3	1	3	3				1			

Notation The ranking method for problem solution has four levels:

3	highly applicable
2	sometimes applicable
1	only rarely applicable
(Blank)	not applicable.

5 Test methods

Section Test method **Page**

Each of these sections has been prepared to a similar format under six subsections, with further sub-headings.

5.X.1 Application and use
Principle, Other names, Application, Usage (Materials, Elements), Use with, Heath and safety, Costs and time, Relevance, Advantages, Prompts and pitfalls, Hints on accuracy.

5.X.2 Principles
Property, Test basis, Measurements, Units and scale, Sensitivity, Accuracy.

5.X.3 Equipment
Components, Size, Weight, Storage, Calibration.

5.X.4 Method of operation
Requirements (Samples, Environmental, Situational, Human resources/skills, Initial trial testing), Use, Data recording, Results, Specification.

5.X.5 Case study(ies)
(Construction, Reason, Results, Success, Assessment).

5.X.6 Relevant documents
Standards and legislation, Guidance, Key reading, Further reading.

Acknowledgements to the author(s) for each of these sections are given in Appendix A and a list of manufacturers (or suppliers) of the test equipment in Appendix B.

5.1 Calcium carbide moisture meter

5.1.1 APPLICATION AND USE

Principle

The calcium carbide moisture meter is a pressure-based test where free water in a sample is converted to acetylene gas by mixing with calcium carbide powder.

Other names

Speedy moisture meter, CM tester, carbide bomb.

Application

The calcium carbide moisture meter provides a site quantitative assessment of moisture content as a percentage of the wet weight of drill extracted material samples. The meter can be used:

- when determining the site moisture content of certain wall and floor materials
- when testing for moisture gradients within the height and/or the thickness of a wall
- as part of the testing regime for the determination of rising dampness, condensation and penetrating dampness
- when testing moisture contents against authoritative references in the consideration of problematic moisture presence and/or the possible need for remedial action
- when verifying/calibrating other moisture measuring devices (see Section 3.3 *Moisture measurements*)
- in determining the moisture content of soils, sands and fine aggregates.

Usage

Materials

Brickwork and blockwork
Mortar and plaster
Cement based screeds and concrete
Soils, sands and fine aggregates.

Elements

Walls
Floors (cement based floor screeds, concrete slabs)
Roofs.

Use with

This meter is generally used in conjunction with electrical resistance or capacitance moisture meters, which are used to identify the locations for sample extraction.

Health and safety

It is essential that the calcium carbide powder should be stored in dry conditions as it evolves a potentially explosive gas when mixed with water. The manufacturer's recommendations should be followed for transportation of the powder, regarding the prevention of contact with the skin or eyes and other safety procedures.

When operating the meter the gas generated in the pressure flask is acetylene, which is flammable. Releasing the gas after a test should be to the external air or in a well-ventilated room. Care should be exercised when operating the equipment externally. External use in wet weather conditions should be avoided.

Clearly no smoking should be permitted during the operation of this test.

Costs and time

The location of the points of extraction of drilled samples for moisture content determination is usually assessed using an electrical resistance or electrical capacitance moisture meter on the surface of the element. The comparative readings from these meters highlight the damp areas, thus preventing random undirected sampling. The time taken will vary considerably with the size of the element and the magnitude of the dampness problem.

It takes approximately 5 minutes to test each sample; it will also take some minutes to extract the drilled sample from the element. In practice this probably equates to 10 minutes per sample tested, in addition to the time required to locate the sampling point.

Relevance

This test method is particularly useful in:

- the analysis of abnormal moisture presences
- situations in which materials other than water may be driven off during oven drying eg a proprietary screed, rich in an admixture, where accurate actual moisture contents are required to assess drying progress
- the assessment of moisture levels within the body of an element which authoritative sources and research experience has shown as vital for the correct diagnosis of wall dampness problems.

Advantages

- A quick site method (minutes).
- Similar accuracy to the laboratory equivalent test (see Section 5.9 *Oven drying method*).
- Allows a number of samples to be tested in one site visit.

Prompts and pitfalls

- Appreciate the value of undertaking a trial assessment of the suitability of the sampling technique as, for example, concrete is usually too hard and too much heat is generated to obtain a representative drilled dust sample.
- Have knowledge of the material under test to ensure the correct interpretation of the results.
- Appreciate that the depth from which the sample was obtained may influence the results. BRE Digest 245, for example, suggests that the first 10 mm of material is discarded and the sampled extracted between 10 mm and 80 mm depth. In damp walls moisture gradients are to be expected within the thickness of the material as well as in the height.
- Ensure that you use a version of the calcium carbide moisture meter that is appropriate for the size and type of material you are testing.
- Screeds containing hydrated calcium sulphate require the use of a version of the carbide meter that can measure up to the 0.5 per cent graduations limit.

Hints on accuracy

- Ensure that the pressure bottle is in calibration and is cleaned each time it is used.
- Endeavour to extract a sample of a single material ie prevent mixing of materials.
- Extract the material sample using a low speed electric drill. Drill for short periods so that heat does not build up and dry out moisture; a 9 mm drill bit is recommended (BRE Digest 245) to minimise evaporation losses through friction.
- Process the sample immediately to prevent the possibility of moisture loss before testing.
- Ensure that the carbide powder is fresh. Powder that is ageing and starting to lose its dark colour should not be used.

5.1.2 PRINCIPLES

Property The calcium carbide moisture meter is a pressure-based test.

Test basis Pressure within a sealed container is created by the generation of acetylene gas from the free water in a fixed sample weight of the material mixing with calcium carbide powder.

Measurements Calibrated pressure gauge directly reading to moisture content.

Units and scale Percentage moisture content (percentage wet weight) generally in the ranges of 0–11 per cent, 0–20 per cent or 0–50 per cent (varies for manufacturer and model).

Sensitivity 0.1–0.5 per cent scale graduations (varies for manufacturer and model).

Accuracy Unlikely to be better than ±1 scale graduation (to be confirmed from calibration).

5.1.3 EQUIPMENT

Components A low-speed electric drill and 9 mm drill bit for sample removal.

Calcium carbide moisture meter comprising mechanical weigh balance, pressure bottle and calcium carbide powder (see Figure 5.1.1).

Airtight containers if removing samples from site to the laboratory.

Size Generally housed in timber box brief case to suitcase size.

Weight The smaller models generally used for wall dampness assessment weigh 5.5 kg, larger models weigh up to 12.5 kg, including box.

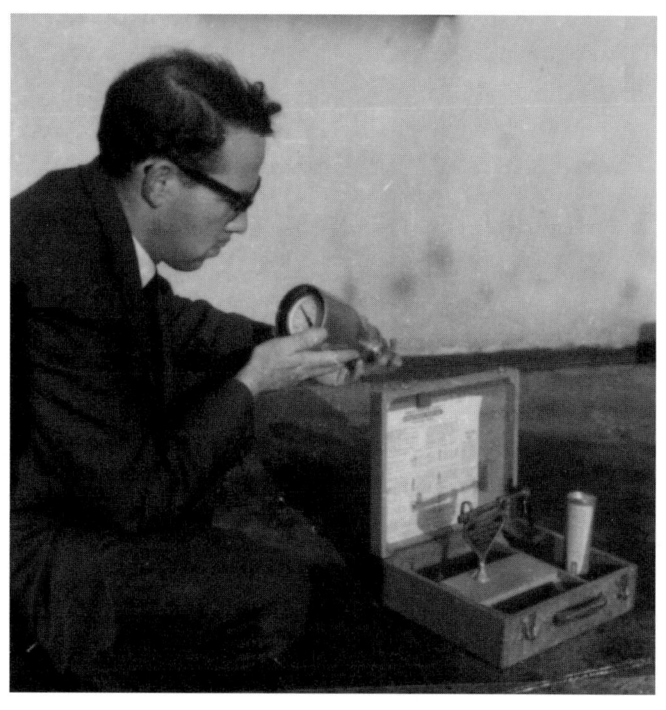

Figure 5.1.1 *Showing the calcium carbide moisture meter*

Storage It is important to store the calcium carbide powder carefully in dry conditions.

Calibration Verification of the accuracy of the readings produced should be undertaken periodically. If it is necessary to recalibrate the gauge, the manufacturer's test and re-calibration service or a calibration rig can be used. Some manufacturers provide a pressure tester for this purpose.

5.1.4 METHOD OF OPERATION

Requirements

Samples Sufficient samples should be tested to confirm or disprove the preliminary findings from electrical meter reading survey.

Environmental The meter is generally used inside a building and a well-ventilated location should be selected for this part of the operation. Rain can contaminate samples if any part of the process is carried out outside the building.

Situational Enough space is required to operate the drill and collect the sample. A level surface is all that is needed for setting up the mechanical balance that is fixed permanently into the carrying case.

Human resources/ skills The user needs to be familiar with the manufacturer's information on the use of the moisture meter and with the contents of authoritative documents such as BRE Digest 245.

Initial trial testing Electrical meters (capacitance or resistance) are often used to identify the locations for samples to be obtained; it is important to appreciate the value of undertaking a trial assessment of the suitability of the sampling technique.

Use Figure 5.1.2 shows the stages in operating a typical calcium carbide moisture meter.

1. Accurately weigh the sample.
2. Carefully place the entire weighed sample in the pressure bottle.
3. Place two full scoops of carbide powder into the top of the pressure bottle.
4. Seal the pressure bottle ensuring that the carbide powder is not allowed to mix with the sample until the bottle is sealed.
5. Upturn the sealed bottle and shake vigorously to ensure complete mixing of sample and carbide powder.
6. Take the reading holding the bottle horizontally with the scale vertical.

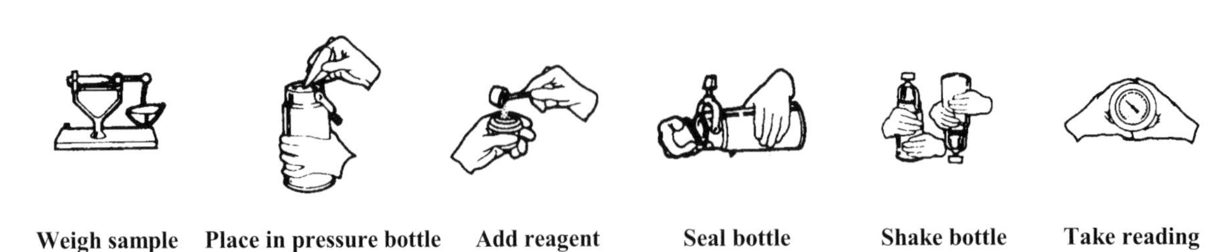

Figure 5.1.2 *Showing the use of calcium carbide moisture meter*

Data recording

Measurements of percentage wet weight moisture content should be recorded in a tabular or graphic format together with all other relevant information (eg date, time, temperature, location) and referenced to any other information taken during the investigation.

Results

The moisture content readings taken with the calcium carbide moisture meter represent the free moisture content present in the tested sample. The free moisture content may be an amalgam of two components: capillary moisture and hygroscopic moisture. The test does not distinguish between these components (see Section 3.3 *Moisture measurements*).

Specification

It is important to identify the material(s) to be tested to enable the correct model(s), in terms of sample size, scale graduations and moisture range to be selected.

For further general guidance see Section 3.4 *Standards and specification.*

5.1.5 CASE STUDY

See Section 5.9 *Oven drying method.*

5.1.6 RELEVANT DOCUMENTS

Standards and legislation

British Standards Institution, *Code of practice for installation of chemical damp-proof courses,* BSI, London, BS 6576: 1985

British Standards Institution, *Specification for the performance of damp-proof courses installed to prevent rising damp,* BSI, London, DD 205: 1991

Deutsches Institut für Normung, *Hot water floor heating systems design and construction,* Beuth Verlag GmbH, Berlin, DIN 4725: Part 4: 1992

Guidance

See information that is provided with manufacturers' equipment, which generally give tables for the interpretation of data.

Key reading

Building Research Establishment, *Rising damp in walls: diagnosis and treatment*, BRE, Garston, Digest 245 minor revisions ed 1986, reprinted 1989

Further reading

Cheetham, D W and Howard, C A, *Translating research into practice – Wall dampness diagnosis – let's get it right*, Building Engineer, February 1999

Howard, C A, *An evaluation of the techniques employed to diagnose rising ground moisture in walls*, Liverpool Polytechnic, MPhil thesis, 1986

5.2 Electrical capacitance meter

5.2.1 APPLICATION AND USE

Principle

Electrical capacitance meters create a harmless electrical field in the material directly beneath transmitter/receiver electrodes contained in or linked to the meter and measure the response. The wetter the material, the greater the response.

Other names

Radio frequency (RF), impedance, dielectric or electromagnetic wave meters permittivity method.

Application

Electrical capacitance meters are suitable for in-situ non-destructive and instant measurement of the sub surface moisture in a variety of building materials. They are used as an alternative to electrical resistance meters (see Section 5.4 *Electrical resistance meter*).

Electrical capacitance meters provide a convenient way for carrying out home and building moisture surveys directly or through a covering or coating. Also, as there is no damage to the element under test, they can be very useful for quality control on the production line or on finished work. Readings are shown on either an analogue or digital dial and some meters are calibrated for specific materials, eg timber, concrete, gypsum, where readings are given in percentage moisture content. Relative readings may be given for other materials such as plaster, brick, block, plasterboard, insulation and roofing materials. Alternatively the manufacturer may provide conversion charts or tables to aid providing data in percentage moisture content terms.

The method has been in use for moisture detection and evaluation for the past 30 years. It has been used as a means of measuring timber and moisture in buildings and damp analysis for 25 years, for roofs for 20 years and as a means of measuring moisture in concrete flooring for past 6 years.

Capacitance meters are typically hand held, pocket size and battery operated, although there are larger instruments available that are fitted with wheels for large area surveying across roofs. There are six generic types of capacitance meter currently available.

- General-purpose meters, for building surveys, with sensitivity ranges for plaster, brick, drywall, timber, roofing felts and insulation.
- Flooring meters with scales specifically calibrated for concrete, gypsum and having a relative scale.
- Flat roofing meters, hand held, or wheeled and for epdm roofing.
- Marine applications with scales for glass fibre, glass-reinforced plastic (grp), wood in fresh water and wood in salt water.
- Highly sensitive meters with signal penetration of up to 75 mm through expanded polystyrene exterior insulation finishes.
- Timber meters specifically calibrated for wood with temperature read out, reading hold facility and an adjustment control for relative density to cope with different species.

Usage

Materials

Brick/block
Cement and gypsum based screeds
Ceramic tiles*
Concrete
Floor covering materials*
GRP
Insulation materials
Paints, varnishes and some other coating materials*
Plaster
Roofing felts and some waterproofing membranes*
Wood and timber components (both as a material prior to assembly or treatment or as a component already part of a building).

* **Note:** These materials do not normally hold moisture; the meter will transmit its detection signal through these materials to measure moisture in the substrate behind it.

Elements

Ceilings
Floors
Roof decks
Walls
Wood and timber components.

Use with

If required, this type of meter may be used in conjunction with calcium carbide moisture meters, humidity sensors, electrical resistance meters or oven drying and when sampling (eg core, lump, drilled) for other tests (eg chemical, strength).

Health and safety

There are no known health hazards associated with the use of electrical capacitance meters although the apparatus produces low power, high frequency electromagnetic current. The meter should comply with CE emission controls and should carry the CE mark.

Care should be taken in gaining access to measuring points eg when climbing ladders and close to unprotected edges of flat roofs, roof lights, etc. Instruments should not be carried on neck straps while on ladders or scaffolding.

Costs and time

This method is fast and easy; readings can be taken literally in seconds.

Relevance

Electrical capacitance meters are used to determine moisture content, and excess moisture in building elements and materials as part of quality control procedures for finished work, both new and remedial, or for problem analysis to:

- determine the suitability of a material to accept a covering or finish
- determine the suitability of timber for installation, fabrication or treatment
- estimate moisture content in walls, roofs, flooring, roof decks, and basements and crawl spaces
- locate, determine and evaluate moisture intrusion into the building envelope and help to trace it to its point of entry
- determine moisture gradients within the plane of a wall, roof or floor

- measure the moisture in timber building components to determine possibility of decay
- determine (by scanning) the directions of moisture movement in building elements
- determine moisture location in built up roofing and tracking leaks
- test* for rising damp, moisture penetration, and condensation in buildings
- test* for termite infestation and other wood and building damaging insects
- test for moisture as part of corrective action
- check for sub surface water leaks
- check for water tightness on a newly finished waterproofing before handover.

* **Important note**: It should be noted the meter in itself would not tell that rising damp, termite infestation, etc is present but potentially provides valuable data. This data should be used to make a better-informed judgement as to the presence and source of the conditions mentioned above (see Section 4 *Diagnostic strategy/procedures*).

Advantages

- Immediate reading in seconds; no waiting or necessity to leave instrument on site.
- Non-destructive; no need to take samples or to damage the material or element being tested in any way or puncture surface.
- Meaningful reading in percentage moisture content for specified material (commonly wood concrete) and relative readings for other materials.
- For timber, readings are not species dependent, but are affected by density and temperature.
- Numerous readings can be taken in one site visit, which allows large areas to be scanned quickly to give a better overall picture of moisture conditions and profile.
- Sensitive up to 25 mm depth but can penetrate to greater depth with some models.
- Some models are pocket-size and battery-powered.
- Compares well with oven drying tests (note: only applies for specific materials where percentage moisture content readings on the meter have been calibrated).
- Readings can be taken through materials, eg tile, roofing felts, and grp, glass, and wall and floor coverings, without damage or puncturing these materials.
- Repeatability of results.
- Not affected by surface contamination or salts.

Prompts and pitfalls

- The signal will not pass through metal material that may be located between the electrodes and the material being tested eg aluminium foil. The presence of metal, eg metal lath, can affect readings.
- If the surface is rough, the readings are likely to be low. In such cases, for "soft" materials, it is helpful to apply some force to minimise the air gaps that affect the readings.
- Density variations within given material substantially affect the readings (eg knots in timber). Where possible tests should be made on representative areas.
- Direction of grain in the case of timber can affect reading; seek manufacturer's recommendations for specific information.
- The meter will give reduced readings for a substrate through a thick coating.
- Elevated readings may be due to contaminants or certain additives.
- The meter response is non-linear with depth (based on inverse square law).

- Some screed additives or residues can affect readings and may give false readings.
- **Beware** the meter may be material specific, ie only give relative readings on materials other than wood. Most material specific meters have at least two scales: one for percentage readings and one for relative readings. (Comparison against oven dying method, for example, may be needed.)
- Electrical meters have been and are still the most widely used meter type for the detection and evaluation of dampness problems as they provide a non-destructive test (NDT) method. The use of these conflicts with the methodology for dampness diagnosis in walls provided by the BRE; BRE suggests the use of electrical meters as preliminary rather than sole instruments in a diagnosis (see Section 3.3 *Moisture measurements*).

Hints on accuracy

- The surface should be smooth, clean and dust free. The electrodes should also be clean, dust free, dry and not damaged or worn through.
- Meters should be pressed firmly against surface so that the entire electrode surface is in contact with the test material (partial contact could result in lower readings).
- Avoid using in presence of temporary surface moisture.
- Check calibration and battery level (this check is built in) before each series of test.
- Measurements should not be made on thin materials as the readings may be distorted by the presence of other materials behind. Also avoid small sections, ie less than the electrode area as these can give distorted readings.
- Where possible, several readings should be taken around the same location.
- On rough surfaces take a number of readings close to one another and if readings vary use highest value.
- Temperature and density affect the readings so accuracy is reduced unless these factors are taken into account, refer to manufacturers' instructions for temperature corrections. Some models have an electronic compensation facility for temperature and density.

5.2.2 PRINCIPLES

Property

Electrical capacitance meters take measurements of electrical impedance. Impedance is an electrical AC measurement combining both resistance and capacitance while at the same time overcoming their separate limitations (ie single line measurement with resistance, and shallow depth of signal penetration with capacitance.) With impedance measurement, two electrodes are fitted at the base of the instrument. A harmless electrical field is set up in an area beneath the footprint of these electrodes from which a low frequency signal is transmitted into the material being tested, measuring the changes in electrical impedance in this material. The wetter the material, the greater the impedance (see Figure 5.2.1).

Test basis

Figure 5.2.2 shows the penetration of the electric field into the material. The electric field penetrates typically to a depth of 20–25 mm, up to 75 mm on some models, so that the presence of water down to such depths influences the field, thus changing the capacitance and affecting the readings. However, the response is inversely proportional to the square of the depth, therefore moisture near or on the surface has a greater effect on the readings.

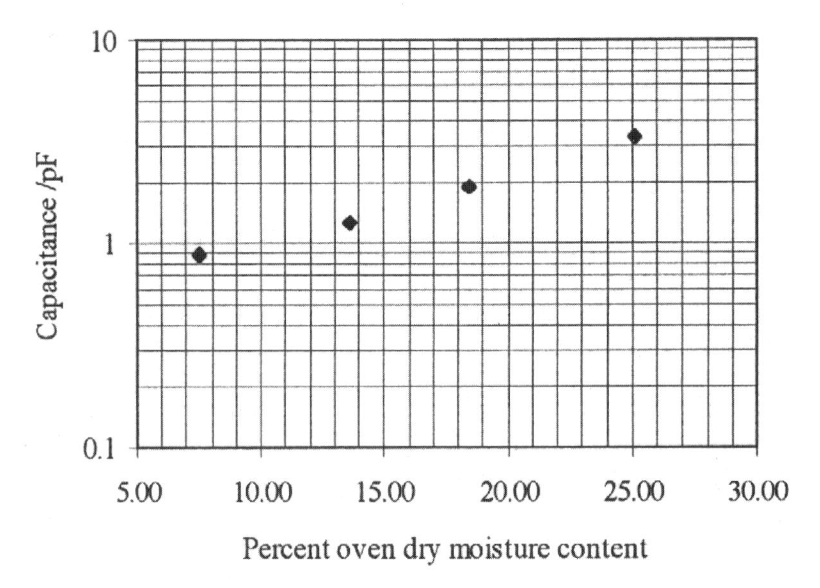

Figure 5.2.1 *Results showing variation of capacitance with wood moisture content (Jazayeri and Ahmet, 1998)*

Measurements

Electrical capacitance meters may be calibrated to measure the moisture content in absolute terms for timber, concrete or gypsum. Other meters may be used for comparative purposes or quantitatively, provided that the user carries out the appropriate calibration or uses appropriate charts or conversion tables.

Units and scale

These instruments are normally calibrated in percentage dry weight moisture content for a specific material and may also have a relative scale for use with other materials. A typical moisture content range for timber is 6–35 per cent.

Sensitivity

Instruments are typically sensitive to changes of ±1 per cent moisture content in timber; analogue meters have divisions in increments of 1 per cent for timber.

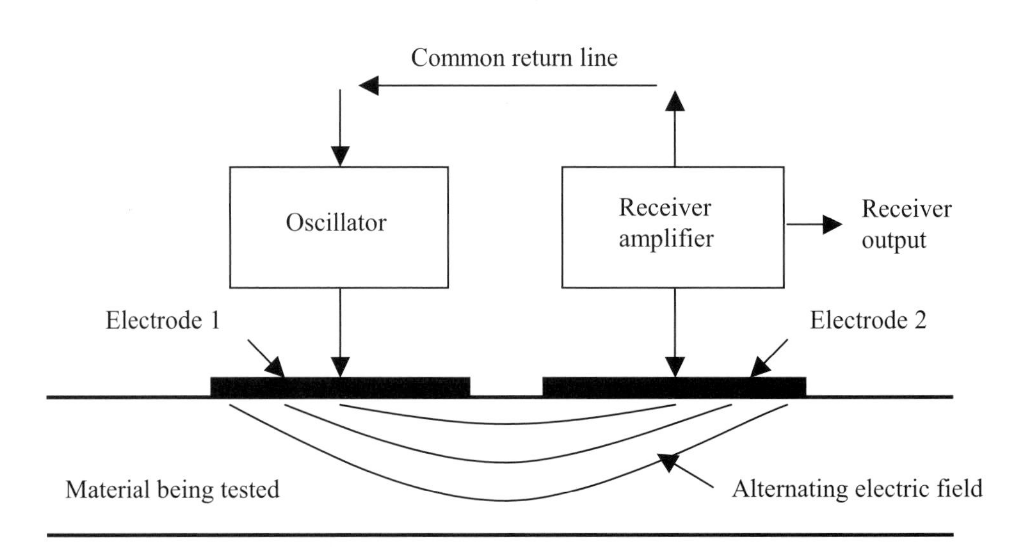

Figure 5.2.2 *Sketch showing the basic principle of operation*

Accuracy Electrical capacitance meters can be accurate to ±2 per cent moisture content in timber in the range 8–25 per cent, assuming that the material is homogeneous and where all required corrections are made.

5.2.3 EQUIPMENT

Components Generally, the electrical capacitance meter is a small hand-held portable instrument with battery in a plastic pouch; see Figures 5.2.3 and 5.2.4.

For roof surveys there are slightly larger hand-held instruments and large wheeled mounted instruments with integral conductive rubber electrode pads to ensure close contact with the substrate. The latter is supplied with case, batteries and removable handle.

Size Small hand-held meters have a contact area 70–100 mm × 150–200 mm and depth 40–100 mm.

For roofs, the hand-held models have a larger contact area up to 125 mm × 275 mm, the wheeled models are larger, typically suitcase-sized.

Weight Small hand-held meters weigh 300–500 g.

For roofs, the hand-held models weigh up to 2 kg, wheeled models weigh 10 kg; 15 kg with the case.

Storage A storage case/bag should be used when transporting or when instrument is not in use. Ensure dry place for storage.

Calibration A built-in calibration/battery check is generally provided. A special external calibration checker is available. If there is any doubt regarding the calibration, the manufacturer should be contacted: an annual calibration is usually recommended.

A calibration table or graph can be produced by taking a series of measurements on samples of a given type of material and then correlating them with, for example, oven-dry moisture contents.

Figure 5.2.3 *Electrical capacitance meter for wood with digital scale*

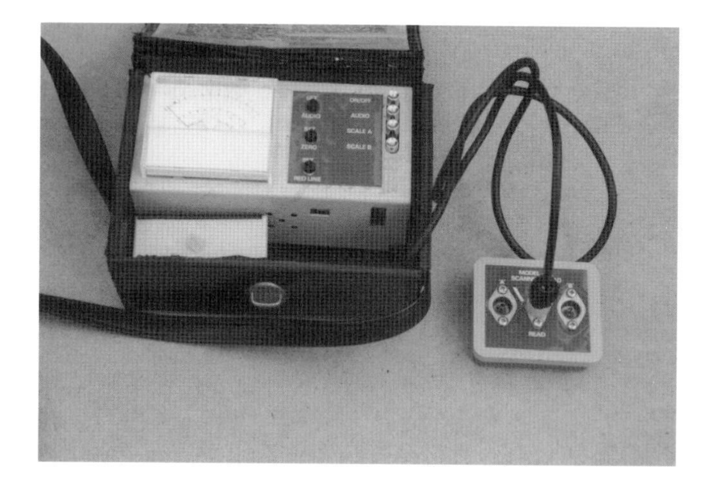

Figure 5.2.4 *Electrical capacitance meter with remote electrodes and analogue scale*

5.2.4 METHOD OF OPERATION

Requirements

Samples Sampling is carried out at appropriate intervals. In surveying the timber components of a room, for example, measurements may be made (say) every 1 m on skirting boards. When conducting a roof survey a convenient grid should be selected, eg 2 m. If there appears to be a problem checks should be made at more frequent intervals. The grid system should be transferred to a sketch of the element being tested and the grid points should be labelled.

Environmental Operation of the electrical capacitance meter is generally from inside a building. When outside, the technique should only be used when an element is surface dry, ie does not have recent moisture on it such as rain or dew.

Situational Where the sensors are built into the instrument, the meter may be difficult to read when used in awkward locations. Some meters have reading hold or freeze and some meters also have an audio warning device that increases in pitch as moisture readings increase.

Human resources/ skills The user needs to be familiar with manufacturers' information on the use of this moisture meter and also with the construction details and some knowledge of the materials being examined. Care must be taken in interpretation of readings, eg if the meter reads 30 per cent moisture content in what appears to be a dry environment then further measurement/ checks should be carried out before the reading can be accepted. Experience in recognising erroneous readings is important.

Initial trial testing Initial trial testing is not necessary, especially for comparative work. However, making measurements on conditioned samples of known density and then oven-drying the samples helps to validate the method. Checks against a calibrator are also useful before commencing tests.

Use

To take a reading: switch on instrument (note that some instruments have battery saver cut-out that switch off after a few minutes of inactivity, a light indicates when instrument is switched on). Press the instrument firmly against the surface making sure it is set at the appropriate sensitivity range, if applicable, for the material being tested. Details of choice of sensitivity should be provided with the meter. Read moisture content from the indicator. On an analogue version, this indicator may also be colour-coded for easy reference and guidance, ie green/yellow/red to indicate low, medium and high moisture levels respectively.

Where there is no electronic temperature/density compensation facility, adjustment should be made to the moisture content reading as directed by the manufacturer. Meter calibration temperature is usually at 20°C.

Data recording The absolute percentage dry weight or relative moisture content is simply read from the analogue or digital indicator on the front of the instrument.

There are electrical capacitance meters available that have microprocessors, non-volatile memory providing data storage and with automatic density, materials and temperature correction. The meter can be interfaced with a PC to download the data for transfer into a word processing or spreadsheet software system.

Results The moisture content reading taken with the electrical capacitance meter usually represents the total moisture content expressed as a percentage that will be absolute for the material for which the meter was calibrated, eg wood, or relative for any other material. The sensitive area is below the electrodes, up to 25 mm deep, depending on material being tested and the model of meter being used. As moisture tends to move upwards and outwards during drying and curing in the case of concrete and timber the reading should be representative of the moisture content lower down from the surface. Coating and covering manufacturers' recommendations should be consulted to obtain moisture levels suitable for safe installation of their respective materials.

There are specialist electrical capacitance meters available that have a deeper signal penetration than indicated above for use on roofs and certain types of wall construction.

The moisture readings may be annotated on roof/floor plans or wall elevation drawings. Readings of equal moisture content may be linked to provide a contour map clearly indicating areas of low and high moisture

Specification For general guidance see Section 3.4 *Standards and specifications.*

Reference should be made to the relevant clause of J42 of National Building Specification when requiring roof surveys.

5.2.5 CASE STUDIES

Large roof As a result of poor workmanship at the time of construction in the early 1990s, insulation was allowed to get wet prior to laying a mastic asphalt covering. A wheeled model electrical capacitance meter revealed large areas of high sub-surface moisture.

Areas of high and low sub-surface moisture were cut out and samples of the insulation analysed to determine the moisture content by the oven drying method. The meter survey was confirmed as an accurate portrayal of the sub-surface moisture.

It should be noted that no specific visual defects with the mastic asphalt were identified that could have contributed to rainwater ingress.

Small roof Rainwater egress that was reported in a single location within the internal accommodation beneath a flat roof with a three-layer felt waterproof membrane. No visual defects to the felt covering were identified. Previous repairs in the immediate area of egress to rectify suspected poor detailing had proved unsuccessful.

A survey with a wheeled model and larger hand held electrical capacitance meter pinpointed the source of rainwater to the top of a perimeter up-stand to a roof light. Rainwater had penetrated an unsealed and unprotected edge some distance from the point of egress. In the area of ingress no defects to the felt covering were detected.

Wall An electrical capacitance meter survey was undertaken to the internal surfaces of a solid brick wall construction where, in the course of a dispute, it was alleged that readily apparent damp walls were a result of cold surface condensation. Notwithstanding an analysis to demonstrate that the conditions for significant condensation to occur were unlikely to exist, it was necessary to identify the source of moisture.

The electrical capacitance meter survey identified clearly that areas of high moisture occurred close to external leaking down pipes, around defective window installations and at floor level where the damp proof course was ineffective.

The use of the electrical capacitance meter was particularly appropriate as it is not affected by surface contamination of salts (unlike electrical resistance meters) and does not pepper the wall coverings with pin-holes.

Floor An electrical capacitance meter survey was undertaken on a concrete ground-bearing slab, finished with thermoplastic tiles, to assist with the identification of suspected rising damp.

The survey identified that sub-surface moisture was present in areas typically expected such as fireplace surrounds, service pipe protrusions, washing machine areas etc. Samples of the substrate were however taken to determine the oven dry moisture content and to validate the use of the electrical capacitance meter.

5.2.6 RELEVANT DOCUMENTS

Standards and legislation

American Standards for Testing and Materials, *Direct moisture content measurement of wood and wood-base materials, Test method for*, ASTM, West Conshohocken, PA, ASTM D 4442: 1992

American Standards for Testing and Materials, *Use and calibration of hand held moisture meters, Test method for*, ASTM, West Conshohocken, PA, ASTM D 4444: 1992

Guidance Some manufacturers provide useful guidance booklets. For example, information sheets on the use of their equipment and guides for use on the different materials encountered in moisture analysis. Most manufacturers supply 'question and answer' sheets and technical help by a telephone back-up service.

National Building Specifications, *Electronic Roof Integrity Test*, NBS Ltd, London, J42 pp 22–23, 1995

Key reading

James, W, *The interaction of electrode design and moisture gradients in dielectric measurements on wood,* Wood and Fiber Science, 18(2), pp 264–275, 1986

Jazayeri, S and Ahmet, K, *Moisture gradient studies in timber by measurement of dielectric parameters,* Proceedings of the 3rd International Symposium in Moisture and Humidity (Volume 2), National Physical Laboratory, pp 179–186, 1998

Torgovnikov, G I, *Dielectric properties of wood and wood-based materials,* Springer-Verlag, New York, 1993

Trechsel, H T, *Moisture control in buildings*, ASTM, West Conshohocken, PA, ASTM Manual series MNL l8, 1994. ASTM code pcn 28-0180094-10

Further reading

Roofing Industry Educational Inst., *Roof moisture surveys current state of technologies*, RIEI. Englewood Colorado USA, 1989

5.3 Electrical earth leakage technique

5.3.1 APPLICATION AND USE

Principle

The electrical earth leakage technique operates on the principle that water passing through leaks in roof or other waterproofing membranes can carry an electric current. Wetting the roof surface and generating an electrical current between the surface and the roof deck, which is earthed, and then tracing the strength and direction of the current can locate any leaks in the roof deck.

Other names

Electronic leak detection, earth loop induction test, electrical conductance technique, electrical flat roof leak detection system, sparks test, geesen system, leaktector test.

Application

The electrical earth leakage technique is used to determine the exact location of leak points in waterproofing membranes. It can also be used to determine the waterproofing integrity of new waterproofing installations prior to hand over.

During the past ten years, the electrical earth leakage technique has become popular in the UK and is now probably the most widely used test method for detecting leaks in membrane roofs, podiums, car parks, tanking and other structural waterproofing applications. It is equally applicable as a quality control test method for new membranes on exposed low slope roofs, car parks, tanking and allied applications.

Usage

Materials

Mastic asphalt
Bitumen roofing felts
Single ply membranes
Liquid applied membranes.

The technique **cannot** be used on:

- metal covered roofs
- metal/faced membranes
- EPDM single ply.

Elements

Flat roofs
Podiums
Basement floors.

Use with

This test method normally stands alone from other tests. However, it can be used as part of an overall test sequence to eliminate, or otherwise, waterproofing membranes from an investigative process to determine the source of leakage into building structures.

Core sampling may be required to confirm roof construction.

Health and safety

The voltages employed are low and the battery-powered equipment does not present a risk of electrocution.

During the test, some wet roof surfaces may present a slip hazard.

Costs and time

One person should be able to test 500–1000 m^2 in a day. Much depends, however, upon the roof finish, obstructions to test, complexity of the roof and, indeed, the number of defects traced, as each has to be marked, recorded and isolated. Heavy coverings of chippings need to be wetted down thoroughly prior to the test commencing.

If lightning conductors are embedded in the chippings, the latter would have to be scraped away adjacent to it. On a day that is windy, sunny or both, it is difficult to keep single ply membranes continuously wet, so the test has to be undertaken in much smaller areas. This tends to restrict and reduce the area for available testing.

Relevance

The technique was first developed to trace leak holes in flat roof waterproofing membranes and because of its pinpoint accuracy it has superseded other techniques, such as flood testing. It can be employed in situations where other techniques would be impossible. These include membranes under paving tiles or inverted roofs, where insulation and ballast do not necessarily have to be totally removed for the test except locally once a leak point is indicated.

The technique can be used as a quality control procedure to check the integrity of a recently completed waterproofing membrane. The method is quicker than flood testing, and does not apply potentially high-superimposed loads. It can then be employed to monitor damage caused by following trades in new construction.

The technique is a useful process in the long term monitoring of roof functionality, eg a roof might be tested every 2–8 years to ensure it retains its integrity and to check for damage.

Advantages

- The technique is non-destructive.
- The technique is less weather dependent than other roof moisture survey techniques, such as thermographic inspection. It works well in the rain, as it requires the surface to be wet. This is of particular importance in countries with a maritime climate.

Prompts and pitfalls

- The technique does not work on all types of membranes; in particular, the membrane being tested must not be a conductor of electricity. For example, black Ethylene Propylene Diene Monomer (EPDM) membranes with a high carbon black content are known to conduct electricity, and the technique will not work on these roofs. Similarly it is not possible to test metal-faced felts (see Table 5.3.1).
- In contrast, the materials below the membrane, notably the insulation and decking, need to be good conductors of electricity. For example, foil-faced insulation boards work well especially if they are mechanically fixed to a metal deck. Timber, ply or strawboard decks are not conductors, whilst woodwool, whether or not screeded, can dry out too rapidly after leakage to remain as a conductor.
- The test equipment and method by which it works are not easy to operate to their full potential. As leak detection tools they can be fairly simple to use but, in unskilled hands, some items may be missed.
- Situations where new roofs overlay earlier constructions can present difficulties because the path of the electric current becomes longer, creating greater electrical resistance and reducing the strength of the current on top of the roof. A reduced current means that the technician needs to be closer to the puncture before a reading can be picked up on the receiver, and thus, there is a greater risk of missing a hole.

- It is essential to check and understand the significance of the materials used in the construction thickness to avoid misuse of the equipment and hence misinterpreting the results. For example, tests on a warm roof may show no signs of leakage due to a vapour control layer preventing the necessary electrical circuit to be made and hence the leak being detected.
- The technique can work on roofs covered with loose-laid stone ballast, concrete paving slabs or walkway pads, although the loose coverings need to be lifted locally to gain access to the membrane surface to locate holes. Any metal resting on or passing through the roof, such as pipe penetrations, handrails, fixed plant or gutter outlets, also offer an earth path and should not be assumed to be an active leak. Instead, to isolate it, the loop of wire should locally encircle the item.

Table 5.3.1 *Indicating where the electrical earth leakage technique will and will not work*

	Membrane	Insulation – method of fixing	Deck
Will work	pvc single-ply EP single-ply bitumen felt asphalt liquid coating	mechanically fixed (metal washers)	metal concrete
Marginal	permeable membranes some aged membranes		timber
Will not work	metal-faced felt EPDM single ply	fully bonded	woodwool

Hints on accuracy

- It is essential that the technician adopt a methodical and unhurried approach, systematically working across all of the roof areas, otherwise holes can be missed.
- The material depth to which the technique works is dependent upon the electrical conductivity of the construction.

5.3.2 PRINCIPLES

Property

With any leak through a membrane, water passes through a hole into the substrate. Water is a good conductor of electricity, whereas most membranes are poor conductors. The electrical earth leakage test technique is used to locate a hole in a membrane; it does not identify the amount of water within a construction.

Test basis

An electrical potential difference is set up typically between the roof surface, which is wetted, and the roof deck, which is earthed. The electrical current will flow directly to the hole. Using an ammeter connected to two probes, the direction of the current can be identified and thus, by moving the probes, any hole can be pinpointed (see Figures 5.3.1 and 5.3.2).

Measurements

Because of the high electrical resistance through the roof, the magnitude of the electrical current is relatively small. However, what is of greater importance than the magnitude of the current is the direction in which it flows, leading the technician to the hole.

Units and scale

The ammeter has a central zero position and half-scale deflection of the order of 0.1 μA.

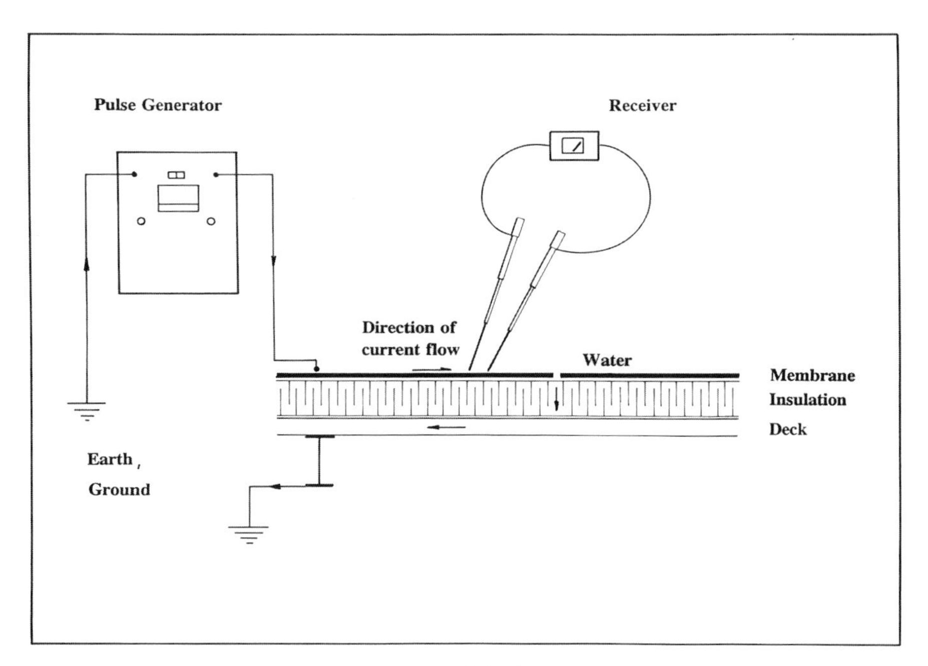

Figure 5.3.1 *Equipment for electrical earth leakage test*

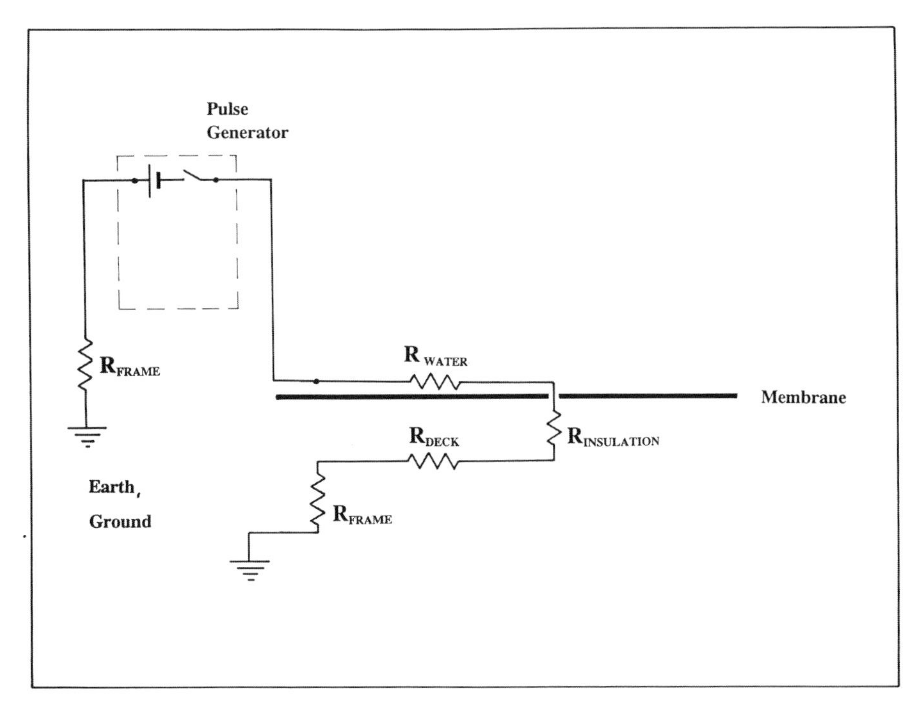

Figure 5.3.2 *Circuit diagram for electrical earth leakage technique*

Sensitivity The sensitivity needs to be such that with very small currents the direction of flow can be readily identified.

The sensitivity of the method in terms of the size of the holes and the distance from which they can be identified is dependent upon the electrical resistance of the construction, as illustrated in the case studies (see Section 5.3.5).

Accuracy Not of any great importance given adequate sensitivity.

5.3.3 EQUIPMENT

Components Items required are an electrical pulse generator, conductor loop, receiver and probes.

The **electric pulse generator**, which weighs approximately 10 kg, can have a rechargeable battery and, when switched on, delivers short bursts of electricity at approximately 40 volt dc potential. The voltage is low so that there is no risk of the technician being electrocuted while operating the equipment neither on the wet roof, nor of affecting the functioning of sensitive electronic equipment within or fixed to the building.

The **conductor loop** is a purpose made wire used to conduct the electricity across wide areas of roofing. With some types of wire small weights are provided to hold the wire in place. This is connected by cable to one terminal of the pulse generator. A separate cable is provided for connecting the other terminal to the earthed deck.

The **receiver** houses an ammeter in a lightweight carrying case supported by a neck strap. **Two metal probes** are connected to the left and right-hand terminals each side of the receiver.

In addition, ancillary items required are hoses, buckets and watering cans, as the membrane surface has to be kept wet throughout test.

Size Generally, all components fit inside a hand-held case.

Weight Approximately 12 kg.

Storage The equipment should be stored in a dry environment in its case. The battery condition should be periodically checked and recharged as necessary.

Calibration As this is a comparative test, calibration is not required, although it is necessary to check the equipment is functioning correctly prior to use and on return to store.

5.3.4 METHOD OF OPERATION

Requirements

Samples It is important that the whole area is surveyed in a methodical and carefully planned way, recording and marking the test area into manageable sized grids. It is sometimes prudent at the start of a test programme to cut a few core samples to check and confirm that the electrical earth leakage points are indeed punctures and that water is present within the roof construction directly below.

Environmental Before setting up the test equipment, it is essential that the surfaces are thoroughly soaked with water. If it is not raining at the time of the test, then water needs to be sprayed from a hosepipe over all of the surfaces to be tested. A surface film of water is required, adequate to conduct electricity. If the roof is sloped and free draining, the water supply needs to be continuous. For new roofs, it is absolutely vital to ensure that any breaches in the membrane have been sufficiently wetted out to allow proper electrical conductance to earth.

Situational Roofs can be dangerous places and appropriate precautions should be taken, especially near unprotected edges.

Human resources/ skills It is important that the technician carrying out the test has been trained and is experienced both with the equipment and construction details.

Initial trial testing Before testing it is essential to check that the roof membrane is a poor electrical conductor and that the insulation and deck are good conductors (see Table 5.3.1 in Section 5.3.1 *Prompts and pitfalls*). It has become normal practice to undertake a site trial of a small area to check the effectiveness of the method on the actual roof construction before progressing to a full survey.

Use

The fully charged electric pulse generator is taken up onto the roof, and one of the terminals is connected to the structure of the building. This could be steel framework, a metal rainwater outlet, or a handrail bolted to the deck. The other terminal is connected to the conductor loop, which is set out in a rectangular shape around the area to be tested. Care has to be taken to ensure that any electrically earthed items within the test area are removed or isolated otherwise they will register as leaks. These would include lightning conductor tapes, roof outlets, and plant resting on the surface. The pulse generator is switched on, a light starts to flash, and a regular clicking noise or a buzzer is heard. The technician with the receiver and two metal probes then steps into the field to be tested and places the two probes onto the roof surface approximately 1 m apart. The zero control is adjusted so that the ammeter is in the central position when there is no current flowing (see Figure 5.3.3).

If there is a hole, the dial will move either to the left or right, indicating the direction of current flow. The technician then moves the probes in the direction indicated, with the strength of the current increasing closer to the hole. By a series of iterations, the direction of current flow shown on the receiver leads the technician to the leakage point. When one of the probes is over the hole, the strongest reading is given. To pinpoint it, one of the probes is rested on the suspect point and the other 25 mm away. At the four compass points all the readings should indicate towards the first probe. At this stage, the technician can normally identify visually some mark or feature on the roof surface, such as scuff or side lap, which can then be ringed in wax crayon for reporting and repair.

Having identified one hole, it is important that the local area is isolated by either laying an isolation loop or by temporarily sealing the puncture with adhesive tape. If this is not done, other smaller holes nearby could remain undetected.

Data recording

It is important to keep written records of the work undertaken and the areas surveyed. All identified leaks should be clearly marked on the roof surface, using a permanent marking system, to enable the necessary repair to be carried out.

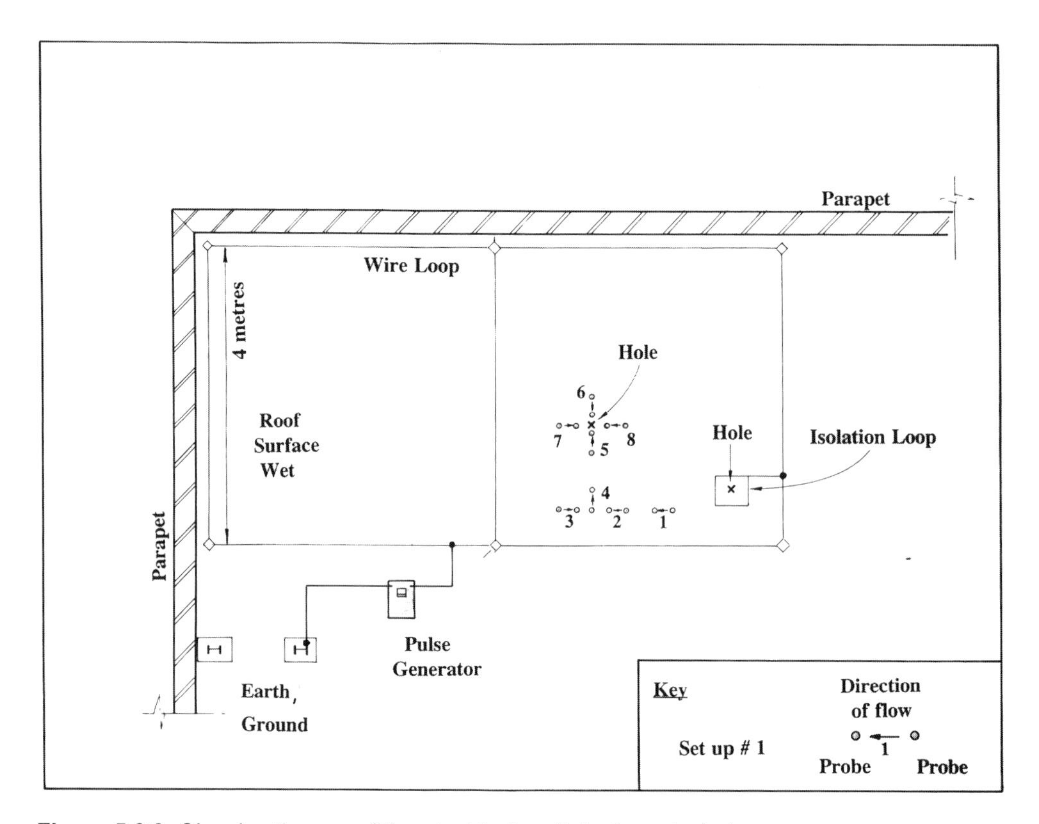

Figure 5.3.3 *Showing the use of the electrical earth leakage technique*

Results

Interpreting the results is probably the most important task of all: to deduce the correct conclusions and decide on an appropriate course of action, to enable specifying the appropriate remedial works to the roof.

As a quality control tool it is essential that an area certified as watertight is actually so and that nothing is missed. Unless leaks in the membrane are correctly identified, isolated and re-tested as necessary, misleading results will be obtained.

Insulated roofs with high electrical resistance pose special problems to the method as no results may be interpreted as indicating a perfect roof, whereas the opposite may be the case. For this reason, the technicians carrying out the testing should have received proper training and be responsible for carrying out a thorough survey.

Specification

Reference should be made to the relevant clause of J42 of National Building Specification. Section 3.4 *Specifications and standards* gives further general guidance.

5.3.5 CASE STUDIES

Study No 1 New single-ply membrane roof

Construction

- White ethylene propylene (EP) based sheet membrane, mechanically fixed
- Polyurethane thermal insulation
- Polyethylene vapour control layer
- Galvanised steel deck.

Reason The single-ply roof membrane had been completed, but before the factory building was handed over other trade contractors started working above the roof without adequate protection and caused puncture damage to the membrane.

Results The technique, shown in Figure 5.3.4, identified the punctures, often hidden below muddy debris. The tests also found local points where the heat welded side laps were not continuous.

Figure 5.3.4 *Showing the use of the electrical earth leakage technique*

Success The technique is a successfully alternative to the physical test where a blunt screwdriver is drawn along the edge of a welded side lap to identify any incomplete welds.

Assessment One of the reasons why the tests were particularly successful in this case study was because the membrane was laid directly over a foil-faced insulation, which, in turn was screw fixed through to the metal deck. Thus, the foil facing acted as an excellent earth, with a strong electrical current flowing over the roof surfaces, enabling the technician to be drawn to the leak point from a distance of up to 3 m.

Study No 2 Over roof

Construction

- Thermoplastic single-ply membrane, fully adhered
- Polyurethane thermal insulation
- Original multi-layer bitumen felt roofing
- Fibreboard
- Channel-reinforced cementitious wood particle decking.

Reason The original built-up felt roof of the residential home had been overlaid with insulation and single-ply membrane. The roof leaked and there was a need to locate all of the holes.

Results None.

Success The testing was not successful.

Assessment The testing was not successful because of the poor electrical conductivity of the decking material. Thus, the probes needed to be very close to the puncture, eg less than 0.3 m, before there was any reading on the receiver.

Study No 3 Aged pvc gutter linings

Construction The construction, see Figure 5.3.5, comprised:

- pvc single-ply membrane, fully adhered
- aluminium support plate
- polyurethane thermal insulation
- coated steel internal face.

Reason Over a period of years, the gutters above a large exhibition hall had been used as walkways, and scuff damage had been caused. The membrane gutter lining had also torn at some of the joints between the aluminium support plates, due to differential thermal movement.

Results The testing, as shown in Figure 5.3.6, identified a large number of punctures, which were later subdivided into two groups – "scuff damage" and "thermal movement" – enabling liability for the cost of the repairs to be allocated.

Success The gutter linings were subsequently repaired and the long-term rainwater leakage problem solved.

Assessment This case study illustrates a successful application of the technique.

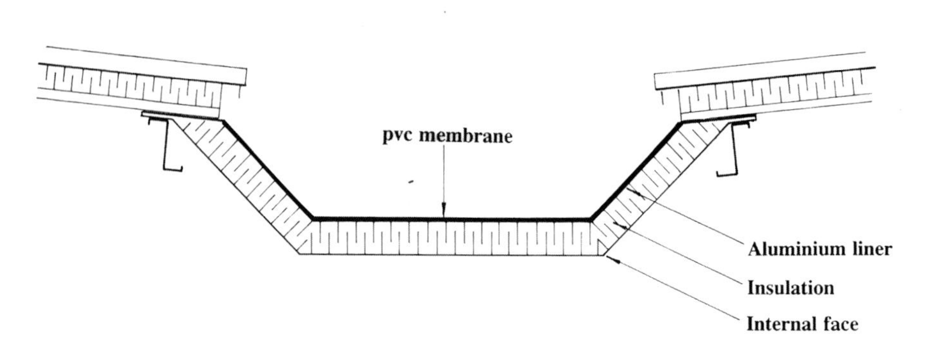

Figure 5.3.5 *Showing the roof construction and gutter detail*

Figure 5.3.6 *Showing the use of the electrical earth leakage technique in gutter*

Study No 4 Aged multi-layer bitumen felt roof

Construction

- Mineral-surfaced multi-layer bitumen felt roofing
- Cork thermal insulation
- Original asphalt roofing
- Lightweight insulating screed saturated with water
- Vapour control layer
- Concrete deck.

Reason As part of a condition survey of an aged built-up felt roof over a hotel, the technique was used to test an area of 300 m^2.

Results It was found that there were 57 holes, notably around penetrations and at the base of a low upstand wall.

Success The technique worked on the vertical faces of the upstand, provided that there was a continuous flow of water.

Assessment One of the lessons learned from the case study was that the technique locates everything, right down to a pin-hole.

It was later discovered that the screed had been wet since the time of original construction twenty years before, when flood testing had resulted in the substrate becoming saturated, which had contributed to the under performance of the roof.

Study No 5 New office complex, London

Construction Building commenced circa 1985, the roofs and podiums were to be covered with insulation and ballast, whilst the latter were being screeded and finished with high-grade walkways, bus stands and roadways.

The waterproofing was a torch applied high performance bitumen felt system applied to a concrete substrate.

Reason 14-day flood testing specified to all waterproofing for roofs and podiums

Results Electrical earth leakage successfully used both with and without flood testing.

Success Following initial trials the requirement for flood test was reduced initially to seven days then later eliminated altogether.

Assessment With certain additional preparatory requirements, the preliminary flood test proved to be unnecessary.

5.3.6 RELEVANT DOCUMENTS

Standards and legislation

No standards are currently available for this type of test.

Guidance

National Building Specifications, *Electronic roof integrity test*, NBS Ltd, London, J42 pp 22–23, 1995

Key reading

Roofing Cladding and Insulation, *Flat roof leak detection: using portable electrical conductance technique*, RCI, London, Technical Note No 35, p 27, Jan/Feb 1994

Roberts, K, *The electrical earth leakage technique for locating holes in roof membranes*, Proceedings of the Fourth International Symposium on Roofing Technology, National Institute of Standards and Testing, Washington, USA, September 1997

5.4 Electrical resistance meter

5.4.1 APPLICATION AND USE

Principle

Electrical resistance meters work by the application of a voltage across two points and the subsequent measurement of the current and hence the electrical resistance. For a given material (species/composition), the relationship between resistance and the moisture content can be accurately established, enabling the moisture content to be measured.

Other names

Electrical resistance meters are also known as conductance moisture meters. This is a slightly more logical name as the electrical conductance increases as the moisture content increases. They may also be called probe or pin meters.

Application

The use of electrical resistance meters is a well-established technique for estimating the moisture content in timber. Correctly used, they are useful for quality control, for example, for checking the moisture content of timber prior to installation into a particular environment (refer to Ahmet et al, 1995).

These moisture meters are also invaluable in comparisons of dampness for other materials provided the same material is used throughout. Oliver (1997) states that the most widely used moisture meter is the resistance-type and that these instruments are useful tools for detecting dampness in masonry and concrete. However, misuse is a problem and Parrett (1997) warns that the "main culprit" of misdiagnosis of rising damp is the moisture meter.

The use, for other materials, of electrical resistance meters calibrated for timber is only advisable when **comparisons** are being made, or when appropriate conversion is made (see also Section 5.4.4, sub-sections *Use* and *Results*); otherwise the values may be meaningless.

Some important points:

- while readings taken on timber are fairly consistent in providing results in the form of percentage moisture content by dry weight, the possible existence of contaminants and surface condensation cause the readings taken on masonry, cement based materials and plaster, to be best regarded as reactions which cannot be quantified in percentage moisture content terms
- it should be recognised that the readings may reflect near surface conditions on elements such as walls and floors unless longer, insulated electrodes are used
- as an essential part of a diagnosis always undertake a visual inspection for defects and faults prior to the use of the meter.

Usage

Materials

Timber
Masonry
Cement-based materials
Plaster.

Elements

Electrical resistance meters are versatile and can be used on materials in all elements, although it may be difficult to read certain instruments where used in awkward locations.

Use with

Electrical resistance meters are usually employed for ease of measurement as an alternative to electrical capacitance meters. They are particularly useful in providing comparative measurements that would generally lead to the need for further testing. This would typically include assessment of actual moisture content and the evaluation of contaminants such as salts. Hence they may be used in conjunction with, for example, calcium carbide moisture meters, humidity sensors and the oven drying method.

Health and safety

Electrical resistance meters are battery powered so there are no electrical hazards.

The electrodes are pointed so there is a small possibility of minor injury during use. Care should be taken in gaining access to measuring points, eg when climbing steps to a loft.

Costs and time

One of the main advantages in use is speed, coupled with the fact that this is effectively a non-destructive test method. Individual measurements can be made in just a few seconds. The time commitment on site will depend on the nature of the inspection; for example a valuation inspection or where a specific problem is alleged.

Contouring of like readings is possible and may be either roughly or accurately undertaken. Often quick visual realisations of the pattern of like readings will assist the diagnosis considerably but again the breadth of experience of the user will be influential.

Relevance

The most common usage of electrical resistance meters is by surveyors checking older properties to:

- confirm the absence of damp
- identify areas of dampness and/or soluble salt concentration in masonry substrates
- test floors, for diagnostic purposes, with the floor covering in situ
- provide an estimate of the moisture content of timber in internal and external environments for quality control or diagnostic purposes, for example skirting boards, floorboards, roof trusses, doorframes.

Electrical resistance meters are also used to locate the positions on the surface of the element that appear to be most problematic and hence direct where further investigation should be undertaken.

Advantages

- Quick, clean and straightforward to use.
- Mainly non-destructive apart from pin-holes from the electrodes.
- Direct readout calibrated to percentage moisture content for timber and considered suitably accurate for practical purposes, typically ±2 per cent (check calibration to verify percentage of dry (or wet) weight and for material).
- Many measurements can be made in short space of time.
- Moisture gradients within an element can be detected (with use of special electrodes on some instruments).

- Allows quick approximate contouring of like readings, which may assist the diagnostic process.
- Light and generally very portable.
- Model with hammer action on the handle to assist in embedding probes into timber (or similar) (see Figure 5.4.6).

Prompts and pitfalls

- It is important to understand that when making comparisons of meter readings the moisture content of different materials in the same conditions varies enormously (see Section 2.5 *Materials and interfaces*). For example, in the same equilibrium conditions the moisture content of timber is 12 per cent, plaster 0.5 per cent and brick 1 per cent.
- It is important to ensure that the correct scale is being read if there is more than one scale on analogue readout meters, or check switch setting on digital readout meters; correction/conversion tables should be used where necessary.
- Most meters will **not** respond when applied to clean, dry, uncontaminated materials, eg brick, mortar, plaster and concrete. However, under moderately humid conditions they will respond to low levels of hygroscopic salt even without a source of moisture. Under the same conditions they will not generally respond to the presence of efflorescent salts such as sodium sulphate. They may not respond either to higher levels of hygroscopic salt contamination where the environment is extremely dry, ie relative humidity less than 40 per cent. For example, there may be no reading if testing inside an airing cupboard. In very wet conditions, the readings may go "off-scale", for example, where a windowsill has been exposed to moisture.
- The readings are usually the result of surface testing only; such readings may not reflect the conditions within the body of that element.
- The presence of surface dampness may greatly exaggerate readings when using meters with conventional, non-insulated, pins. To check whether surface moisture/condensation is distorting readings, first insert electrodes slightly (say to 1 mm depth) and take reading. Then push pins into required depth; the meter should indicate a slightly higher reading (1–2 per cent in timber).
- Testing through conductive materials (eg aluminium foil, conductive wallpaper paste) will give invalid results.
- Pin-holes from the electrodes or small drill holes used to obtain reading at depth may be unacceptable where appearance is important.

Timber

- In timber precision measurement is difficult as preservatives, species (often unknown), consistency and temperature can all affect the reading.

Masonry and concrete

- Most meters are not generally calibrated for masonry, concrete substrates and other hydraulic setting materials. It is best to therefore view the readings as reaction readings, since the meter responds to several parameters in addition to dampness including the inherent variability of material constitution and any impurities present. These are known to distort the readings and prevent consistent correlation with material moisture content values.
- Low readings may be encountered where surface drying has occurred with a dense material, or at the joints in a floor covering.

Hints on accuracy

Readings should be made soon after insertion of the pins and immediately after the instrument is switched on.

- Where possible, several readings should be taken around same location. The average value should then be calculated.
- The instrument or leads must not be allowed to get damp, as the circuitry is extremely sensitive, otherwise false readings may result.
- For surface or near-surface measurements short electrodes may be employed. For establishing the moisture content deep inside a material or at a given depth, long insulated electrodes should be used, see Figure 5.4.7.
- Pins must be inserted to provide tight contact with the material.
- Damaged pins should not be used.
- Ensure that the correct type of pin is used.
- Some instruments give finite readings (eg 5 per cent) even when the pins are not connected. When an actual test yields such low values, check that there is no problem with the connecting lead before accepting the measurement as valid.
- Avoid using instruments in very dry atmospheric conditions, eg a frosty day in winter, as static build-up on the meter may cause wildly fluctuating readings. (This problem may occur with both analogue and digital reading instruments.)
- Routine checking of battery status and calibration are important to ensure reliability.
- For accurate results the temperature needs to be recorded and meter readings corrected accordingly.

Timber

- For optimum accuracy the electrodes must be inserted either parallel to, or across the grain, depending on the calibration orientation. Errors of up to 2 per cent moisture content can result if an incorrect orientation is used.
- If the species is known improved accuracy results from using species correction tables (note that the instrument may have a facility for species correction).

5.4.2 PRINCIPLES

Property

This method tests the electrical conductivity of the material between the electrode pins of the instrument. The readings depend on the resistance to electrical conduction. For all building materials the resistance decreases as the moisture content increases.

Figure 5.4.1 shows the variation of resistance with moisture content for many types of timber. The broad band explains why species corrections are necessary for accurate results. Also, the flattening off at higher moisture contents indicates that these meters are relatively insensitive at high moisture levels.

Note also that at very low moisture contents the material resistance approaches the open-circuit resistance, making quantitative measurements extremely difficult.

It is useful to remember that despite wide calibration ranges, moisture meters have the best accuracy for timber between approximately 6 per cent and 30 per cent moisture content.

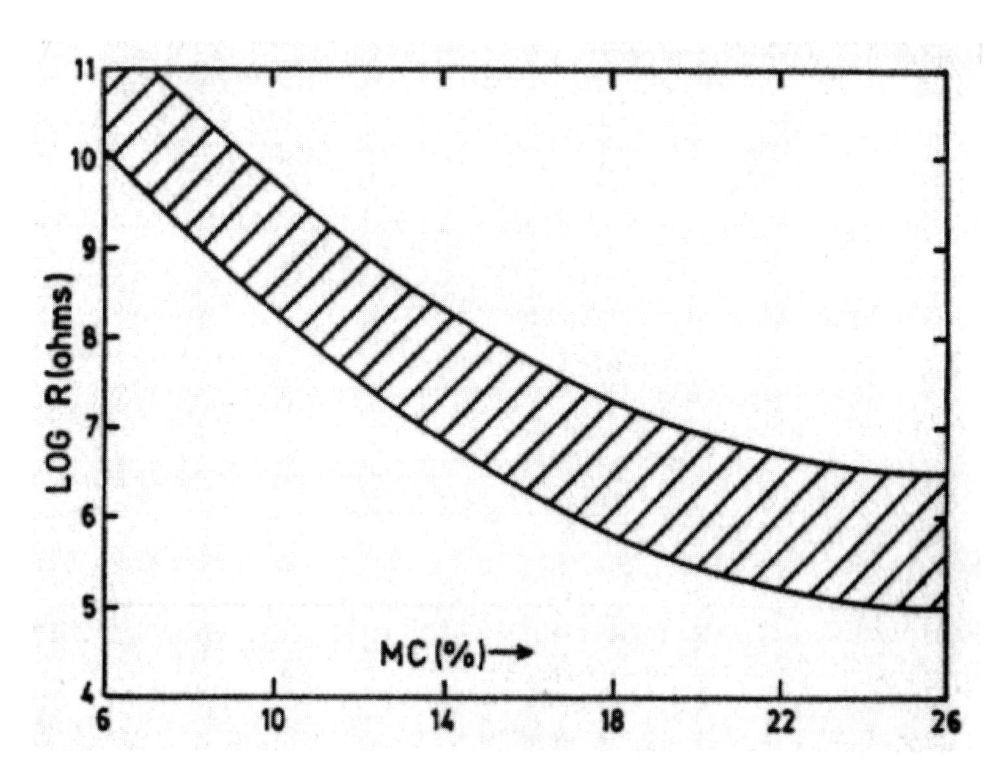

Figure 5.4.1 *Showing the variation of resistance with moisture content for a range of timber species (extracted from Skaar, 1988)*

Test basis

Figure 5.4.2 is a schematic showing the principles of operation of typical resistance-type moisture meter.

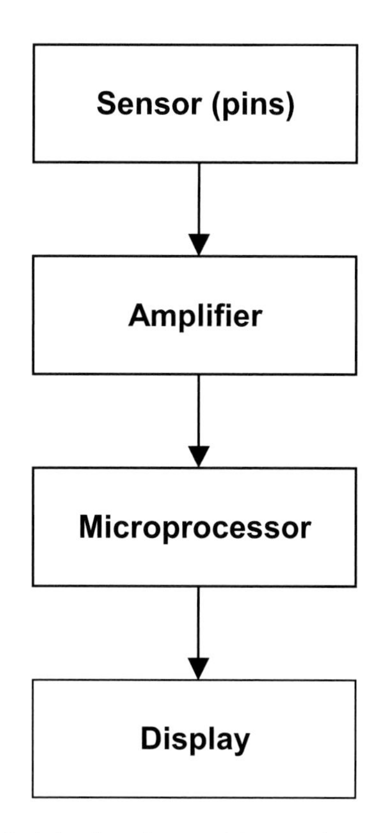

Figure 5.4.2 *Principle of modern moisture meter*

Measurements

The testing device is essentially an ohmmeter, ie resistance measuring device calibrated using relationships similar to Figure 5.4.1 to read percentage moisture content. Materials whose characteristic curve is similar are given the same calibration.

Units and scale

The equipment is normally calibrated in percent dry weight moisture content for timber.

The limits are typically in the range 6–40 per cent moisture content for timber although some meters are calibrated up to 100 per cent.

For use on masonry products, screeds and concrete, there is no effective measurement unit. The equipment may have a separate relative scale for these materials.

The display may be LED, progressive illumination of lights, or traditionally, a needle traversing a scale.

Sensitivity

On digital instruments the scale is usually in 0.1 per cent graduations. Analogue meters usually have calibrations in intervals of ±0.5 per cent or ±1 per cent moisture content.

Accuracy

For **timber**, the best accuracy is about ±0.5 per cent. Generally, in buildings the species will not be known so absolute accuracy may be limited to around ±2 per cent. Applied to timber and where compensations are made for temperature and species, reasonable accuracy can be obtained (see Ahmet et al, 1997).

Figure 5.4.3 shows the comparison of readings taken from a certain moisture meter against oven-dry values during a particular study (see also Section 5.4.5 *Case study – timber*). In this example, the mean is offset from the oven-dry values by about 1.4 per cent moisture content with standard deviation approximately 0.7 per cent. In other words, the calibration accuracy is slightly erroneous. However, the readings are closely grouped around the "true" value. Note that not all meters have such good repeatability, nor do they necessarily read slightly on the low side.

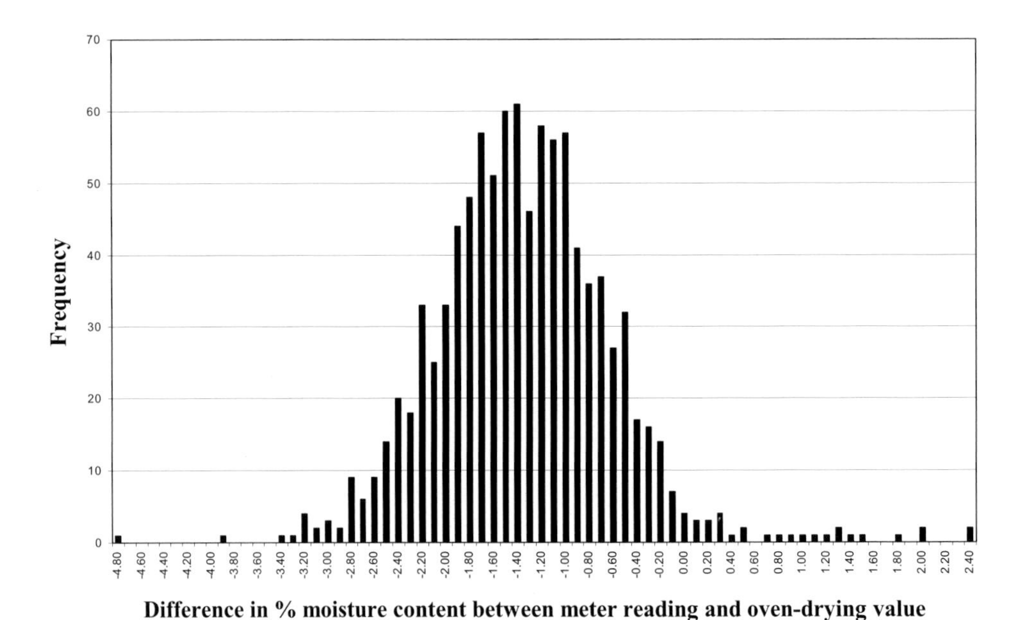

Note: temperature and species corrections have been applied

Figure 5.4.3 *Histogram showing comparison of timber moisture content by resistance meter and oven-dry method at 20°C*

Where used on **masonry, screeds and concrete**, accuracy is a property that is difficult to evaluate as the readings are not absolute moisture content and as so many factors can affect the magnitude of the reading taken. (Sources of possible error are given in Section 5.4.1 *Prompts and pitfalls*).

5.4.3 EQUIPMENT

Components

In some cases the equipment consists of a self-contained meter with the pins directly attached, see Figure 5.4.4. More frequently, the electrodes are remote; being connected to the meter via shielded cable, see Figure 5.4.5. Figure 5.4.6 illustrates a meter where the electrodes are inserted using a hammer action on the handle and Figure 5.4.7 a meter employing long insulated electrodes

Figure 5.4.4 *Compact moisture meter including pins. A socket is included for remote electrodes*

Figure 5.4.5 *Moisture meter with remote electrodes*

Figure 5.4.6 *Moisture meter with hammer-action electrodes*

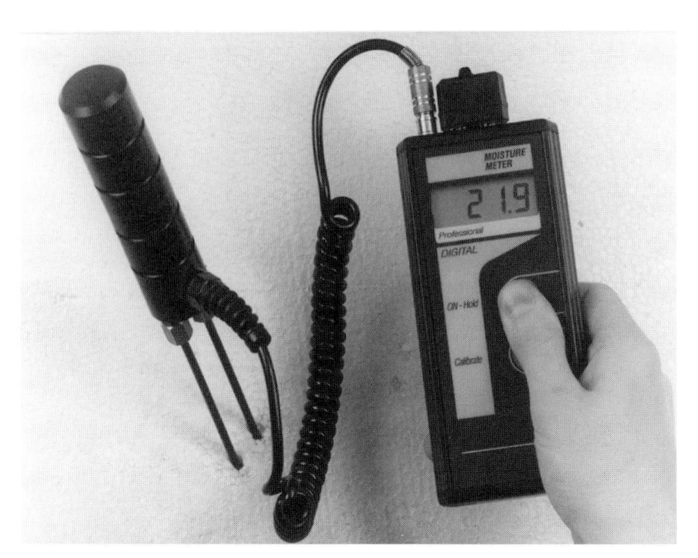

Figure 5.4.7 *Moisture meter with insulated pins for measuring moisture content at required depths*

Size Most resistance-type moisture meters are designed to be hand-held; typical dimensions are 180 × 110 × 60 mm. Where accessories are necessary, meters are supplied in small carrying cases.

Weight Modern meters have mass typically between 300 g and 700 g; hammer-action electrodes can add up to approximately 2 kg.

Storage It is advisable to carry and store the meter in the normally padded box/bag provided by the manufacturer. Contact with moisture and damp materials must be avoided except at the pin end of the electrodes.

Calibration Meters normally have internal calibration check facilities or special resistance boxes may be purchased for external calibration. If there is any doubt regarding the calibration, the manufacturer should be consulted.

5.4.4 METHOD OF OPERATION

Requirements

Samples Sampling is carried out at appropriate intervals. In surveying the timber components of a room, for example, measurements may be made every 1 m on skirting boards. If there appears to be a problem, measurements should then be carried out at more frequent intervals.

On masonry and concrete walls, they are best used by plotting a vertical series of meter readings (**a moisture meter profile**), the pattern of which may give a preliminary indication as to the potential problem source of moisture (Coleman, 1997). Additionally, horizontal readings are typically taken along the wall at regular intervals. These should not only be at floor level but also extend to a height within the reach of the surveyor.

Environmental The equipment must be kept dry at all times. If an instrument stored in a cold environment is to be used in warm surroundings condensation is likely. In such cases sufficient time must be allowed for the equipment to attain ambient temperature before use.

The presence of surface dampness (eg condensation, dew and rain) is likely to yield false readings, caused by "shorting".

Situational It may be difficult to read certain instruments where used in awkward locations, eg in corner of loft where the roof pitch is shallow. This situation is invariably better where remote electrodes are employed.

Human resources/ skills A high level of expertise is not required for using the equipment. However, some basic training and involvement with trials is necessary. Care must be taken in interpretation. For example, if the meter is reading only 5 per cent (timber moisture content) in what is clearly a damp environment then the meter and leads should be checked.

Initial trial testing Initial trial testing is not necessary, especially for comparative work. However, making measurements on conditioned samples of appropriate species and then oven-drying the samples helps to validate the method.

Use

Connect electrodes and leads to the meter. Insert the electrodes into material to be tested and then switch on meter and take reading, see also Section 5.4.1 *Prompts and pitfalls* and *Hints on accuracy*. Valuable information on the use of moisture meters for timber is available in BRE Digest 245 and TRADA *Wood information*, Section 4 Sheet 18.

Data recording

Instruments generally display moisture content readings for timber and it is necessary to log the readings on paper to assist their interpretation. Where there is no automatic temperature compensation, adjustment should be made as directed by the manufacturer where temperatures differ by more than a few degrees from the calibration temperature (usually 20°C). Corrections can also be made for different species and materials. The extent of recording of readings will be inevitably determined by the nature of testing undertaken.

With certain instruments the data can be stored and then downloaded into a computer, which with appropriate software can be used to produce contours of like readings to aid interpretation.

Results Provided that the necessary precautions are taken, the readings should be accepted as valid although there will inevitably be some variations due to statistical fluctuations (see Section 5.4.5 *Case study*). Causes of false readings have been given in Section 5.4.1 *Prompts and pitfalls* and *Hints on accuracy* and precautions should be taken accordingly to minimise these.

Timber always tends towards equilibrium moisture content with the humidity of the surrounding atmosphere. At room temperature the relative humidity needs to be continuously at least 90 per cent or above before timber decay is likely. Moisture meter readings less than about 7 per cent moisture content represents very dry timber. The interval 7–11 per cent is typical of most indoor environments, while unheated buildings can have normal timber moisture contents up to typically 16 per cent. Internally, it is unlikely that the moisture content is above this latter value unless the timber is in contact with, for example, rising damp, rainwater and water from leaking pipes. Rotting is most likely when the moisture content exceeds typically 20–25 per cent. Because of the hygroscopic nature of timber, the measured moisture content is indicative of the extent of dampness in surrounding materials.

On **masonry, cement-based screeds and concrete** the readings taken with this type of meter are generally relative ie not direct readings of percentage moisture content. Some meters give readings of Wood Moisture Equivalent (WME) (see Section 3.3 *Moisture measurements*).

Specification Section 3.4 *Specifications and standards* gives general guidance.

5.4.5 CASE STUDY

Timber

Construction A project jointly carried out by the University of Luton and TRADA Technology Ltd to establish the equilibrium moisture content of timber in a wide range of internal conditions involved the use of moisture meters to measure the moisture content of both standardised samples and in-situ timber.

Reason To establish the range of normal moisture contents in internal buildings for various categories including:

- unheated buildings
- heated domestic
- heated commercial properties (refer to Ahmet et al, 1996).

Results Thousands of measurements were made in these experiments. As part of the study, a comparison between the meter readings and oven-dry measurements showed that for certain meters the species correction factor could be erroneous, producing systematic errors of typically several per cent. Where this was not a problem a normal distribution was produced centred fairly close to the oven-dry value, see Figure 5.4.3.

Success and assessment The data showed that the use of moisture meters is generally reliable for estimating timber moisture content provided that the various precautions are followed. Because of the statistical variations it is advisable to use the mean of at least a few measurements, rather than accepting individual values.

5.4.6 RELEVANT DOCUMENTS

Standards and legislation

At present there are no national or international standards for resistance-type moisture meter calibration (Ahmet, Jazayeri and Hall, 1996). However, there is a proposed standard for the use of moisture meters for timber:

British Standards Institution, *Code of practice for installation of chemical damp-proof courses*, BSI, London, BS 6576: 1985
(Note: The BS is out of date and is likely to be revised using BWPDA Code as a model.)

British Wood Preserving and Damp-proofing Association, *Code of practice for installation of chemical damp-proof courses*, BWPDA, London, 1995

Comité Européen de Normalisation, *Round and sawn timber: method of measurement of moisture content*, Office for Official Publications of the European Communities, Luxembourg, CEN175-13.01: March 1995

Guidance

Timber Research and Development Association, *Moisture meters for wood*, wood information, TRADA, High Wycombe, Section 4, Sheet 18, revised October 1991

Timber Research and Development Association, *Moisture in timber*, wood information, TRADA, High Wycombe, Section 4, Sheet 14, April 1999

Building Research Establishment, *Rising damp in walls: diagnosis and treatment*, BRE, Watford, Digest 245, minor revisions ed 1986, reprinted 1989

Ahmet K, Dai G, Jazayeri S and Tomlin R, *Tests on the use of timber moisture meters*, Journal of the Association of Building Engineers, **72**(6), pp 10–13, 1997

Key reading

Ahmet, K, Jazayeri, S and Hall, G, *Standardisation of conductance-type timber moisture meters*, Proceedings of the seventh international conference on the durability of building materials and components, Stockholm, edited by C Sjostrom, E & FN Spon, London, pp 673–682, 1996, ISBN 0 419 20690 6

Parrett, M, *Managing disrepair in local authority housing: the misdiagnosis of rising damp using electronic moisture meters*, Professional paper 97009, Lewisham Council, 1997

Coleman, G R, *Use of electrical moisture meters*, Building Engineer, 1997

Howard, C A, *An evaluation of the techniques employed to diagnose rising ground moisture in walls*, Liverpool Polytechnic, MPhil thesis, 1986

Further reading

Ahmet, K et al, *The moisture content of internal timber*, Journal of the Association of Building Engineers, **70**(2), pp 18–20, 1995

Ahmet, K et al, *The moisture content of internal timber: 2*, Journal of the Association of Building Engineers, **71**(3), pp 10–14, 1996

Oliver, A, *Dampness in Buildings* (2nd ed), Blackwell Scientific, 1997

Skaar, C, *Wood-water relations*, Springer-Verlag, New York, 209, 1988

5.5 Humidity sensors

5.5.1 APPLICATION AND USE

Principle

The relative humidity (rh) of an enclosed air space formed in, on, or containing a porous material, eg concrete, can be related to the amount of moisture in that material provided enough time has passed for equilibrium to be reached, which may take several days. Humidity sensors provide a measure of the rh of the air space, hence a comparative measure of the moisture condition of many building materials.

Other names

All relative humidity measuring devices may be termed hygrometers.

There are four forms of humidity sensor.

1. **Mechanical hygrometers** that have a clock face and are used for surface applications. There are several types of mechanical hygrometer available, paper, natural hair and synthetic fibre.
2. **Electronic rh probes**, which can be used with a hand held meter or installed as sensors for automated long-term monitoring applications. Electronic rh probes may be termed thermo-hygrometers.
3. Dew-point sensors, which can have a high accuracy and stability but are seldom used for this type of application and hence have not been covered in this section.
4. Hygroscopic cores can be either periodically removed and their moisture condition determined, eg using an electrical resistance meter/by oven-drying or provided with embedded electrodes for automated long-term monitoring. Hygroscopic cores include **wooden plugs** (also termed the Equilibriating Wooden Plug Method) and a special controlled composition of cement paste plugs developed by Parrott (1988) of BCA. The latter are not covered in this section, but operate on the same principles as wooden plugs.

Application

Prior to the application of any impervious finish it is essential to measure the moisture condition of the substrate and to assess the likelihood of further access to any source of moisture in order to minimise potential defects, eg peeling, blistering, delamination. For such measurement, humidity sensors provide the basis for portable non-destructive (surface) and semi-destructive (embedded) test methods.

BS Codes of Practice recommend the use of humidity sensors for the measurement of moisture condition on all newly constructed cementitious bases and screeds prior to installing moisture sensitive floorings. These include adhesive fixed flexible finishes such as vinyl, rubber, carpet, and timber flooring systems. Measurement is necessary on all new substrates whether they are ground-supported floors or suspended floors; moisture is retained for a considerable time and it diffuses slowly depending on drying conditions. There is no necessity to test suspended floors on old properties unless there is reason to believe that the substrates have not dried out, eg through leakage, but ground supported floors should always be checked as they may have a leaking membrane, or worse, none at all. BS Codes of Practice applications have been indicated in the remainder of this section in text boxes entitled **Flooring substrates**.

In situations where surface applied damp proof membrane (dpm) systems or special adhesives with dpm properties are to be used to suppress residual moisture and protect the flooring, the need for measurement may be reduced. Nevertheless, measurement will still be necessary if the manufacturer of the system specifies a maximum moisture condition limit.

Humidity sensors can be used in cavities in construction. They can also be used to determine the effectiveness of cladding, coating and ventilation in reducing moisture levels in construction materials and to check for condensation or high humidity in wall and roof systems and in basement construction.

Mechanical hygrometers have been used for many decades and are very simple to use. Electronic rh probes are used extensively for monitoring ambient rh (and temperature) conditions. They are small easy to use with a battery operated hand-held meter that displays rh (and temperature). They have been applied to the measurement of moisture in the building fabric and are specified, with mechanical hygrometers, in BS Codes of Practice for flooring substrates.

Hygroscopic cores, wooden plugs mainly discussed in this section, were initially used for the long-term measurement of generally high rh (75–100 per cent) in concrete, so that deterioration rates/risks from frost, corrosion, alkali aggregate reaction (AAR) could be determined. Electronic rh probes can be used but the manufacturer's speci-fication should be checked, as many tend to be less reliable at high relative humidity.

Usage

Materials Any porous materials in or on which a sealed airspace can be created, for example:

- concrete
- cement/sand or concrete screeds
- surface smoothing screeds
- terrazzo
- concrete toppings and generally all cementitious substrates
- brick and block work
- plaster
- mineral wool
- wood and wood products
- stone
- sand
- aggregates and soils.

Elements

Floors
Columns
Slabs
Beams
Walls
Cavities in wall, roof and basement construction.

Use with

When positioning humidity sensors, eg on or within a substrate before applying an impervious covering, an electrical capacitance or resistance meter can be used to locate the wettest areas.

The humidity sensors may form part of an overall automated monitoring system, including strain and crack movement gauges, on structures with AAR, for example (see Section 6 *Automated long-term monitoring*). Holes drilled for powder samples for chloride or other analysis can be utilised as locations for these sensors.

Health and safety

Care must be taken in placing the equipment and any cabling to ensure that they do not create trip hazards or obstruct escape routes. Adequate protection and warning signs should be provided. Section 3.5 *Health and safety* gives further general guidance.

Costs and time

Flooring substrates

Make allowance for the time for the test area to achieve equilibrium; typically four days minimum; tests on thick constructions can take two weeks or more.

Make adequate provision for the time and costs for regular site visits to collect the test data, which is generally manually recorded, and to move the equipment, if required, to a new area once equilibrium has been confirmed. Also make provision for regular calibration/verification of the equipment.

Make provision, for all but the smallest projects, for the cost of more than one set of test apparatus to be available for the effective monitoring of moisture ahead of the floor finish laying process.

Make allowance for the time (which may be days or weeks) required for the humidity sensor to reach equilibrium with the material.

Make adequate provisions for the costs associated with data collection: time costs for regular site visits against the capital costs of automated monitoring equipment.

Make adequate provision for the time and costs for regular calibration or verification of all types of humidity sensor.

Electronic rh probes are at least twice the cost of the mechanical hygrometers, and require a data logger and/or meter to enable the readings to be taken.

Wooden plugs can be very inexpensive. The costs are mainly in the time making up the sensor system and, depending on the application requirements, for calibration and verification. They also require a data logger and/or meter to enable on-site readings to be taken.

Relevance

Flooring substrate

BS Codes of Practice recommend the use of humidity sensors to assess whether the moisture condition of bases and screeds is below a specified value or that recommended by the flooring product manufacturer. Construction moisture remaining in concrete bases and floor screeds can, if in sufficient quantity and under certain conditions, adversely affect moisture sensitive floor finishes and the bond and integrity of flooring adhesives. To reduce the risk of failure of such floor finishes it is necessary to allow the residual moisture condition of bases and screeds to dry down to acceptable levels prior to laying finishes.

Deterioration processes are primarily controlled by temperature and rh levels, as distinct from moisture content. Much of the emphasis in moisture measurement in buildings is focused on the less than 75 per cent rh levels where deterioration is stopped.

However long-term data in the greater than 75 per cent rh range, in which deterioration proceeds, is of interest to help in prediction of deterioration rates and in the development procedures for the analysis and prediction of durability design life.

The moisture in the specific material affects not only the material itself but also other materials that come into contact with it. The underlying principle of the rh measurement method is that both temperature and vapour pressure reach equilibrium in:

- the porous material
- the air within a cavity in the material
- within the humidity sensor.

Hence rh data taken in cavities in construction and in sealed drilled holes in porous materials is an appropriate measure to assess long-term building performance or the risk of premature failure.

Advantages

Flooring substrates

Mechanical hygrometers and **electronic rh probes** used for checking floor surfaces prior to laying finishes as specified in BS Codes of Practice have the advantages that:

- it is a non-destructive test
- the rh data provides a qualitative and quantitative assessment
- the test arrangement attempts to replicate the conditions which are likely to be created beneath a floor covering after installation
- **mechanical hygrometers** are cheap, give a direct read out of rh and are simple to use (however, it is also easy to misuse them)
- **electronic rh probes** are easy to use, usually have a digital read-out, built in temperature sensor, remain in calibration for reasonable periods (months) and are reasonably robust.

Electronic rh probes and **wooden plugs** can be used for other applications with the following additional advantages:

- both can be inserted at various depths in the construction thickness to obtain a humidity profile by adjusting the depth and inclination of hole, isolating and sealing the air space containing the sensor and sealing the hole at the surface
- **wooden plugs** can be very cheap, very small and can have good long-term stability.

Prompts and pitfalls

Flooring substrates

Always ascertain the date when the substrate was laid. As a rule of thumb it is recognised that 50 mm thick isolated sand/cement screed will take 50 to 60 days to dry under ideal conditions. At a thickness over 50 mm the relative time increases and it has been known for a 75 mm screed to take 12 months to reach the required level of 75 per cent rh. Power floated or trowelled dense concrete of 150 mm thickness can take years to reach 75 per cent rh.

Adequate time must be allowed for equilibrium to be reached, for tests on thick constructions, this could take two weeks or more.

Mechanical hygrometers are sensitive instruments and are prone to mechanical damage and loss of calibration. Re-calibration prior to each placement of the instrument is essential. Mechanical hygrometers are easy to use but even easier to abuse.

Electronic rh probes require to have a recent calibration/verification, seek manufacturer's recommendations.

The BS Codes of Practice test method is not suitable over surface applied moisture membranes or most proprietary polymer modified screeds.

When locating **electronic rh probes** or **wooden plugs** for other applications, eg within a hole, the following points should be taken into consideration:

- the main difficulties can arise from poor location and sealing of the hole. The hole should be inclined slightly to prevent water running in. Do not create cold-spots by using metal components, eg in expanding bungs, replace with nylon to avoid introducing condensation problems
- if in-depth readings are taken, the holes must be drilled, brushed, vacuumed out and sealed for a period of more than 48 hours in order for the moisture within the material to reach a balance with the air within the hole. With a dense material the period of time should be extended to approximately 72 hours. Confirm that equilibrium has been reached by taking readings over a period of several days (not hours). See additional prompts and pitfalls for wooden plugs, below
- a period of at least 48 hours is necessary if surface humidity readings are taken. However the period required for equilibrium measurement may be longer in certain instances. The readings rely on the speed of evaporation from the material to the surface that is affected by the materials density and any coating previously applied. Note: any coating previously applied must be removed before surface readings can be taken
- the pocket of air in which the rh measurement is to be made should be isolated from the atmosphere by thermal insulation and a vapour barrier ie the air should be at the same temperature as the material under consideration and not be directly influenced by atmospheric moisture. Different types and thickness of material will take different times to reach equilibrium.

In particular when using **wooden plugs**:

- do not re-use after oven drying
- drill an oversize hole, ie 14 mm for a 10 mm diameter plug, to provide adequate space to permit swelling of the plug in the event of very wet conditions
- if the plugs go mouldy there is an obvious problem, the rh is very high and calibration will have shifted
- do not expect plugs to reach equilibrium in a day or two; before fitting they will generally have moisture contents of 8–10 per cent (if stored at 40 per cent rh) and may take a month or more at 95 per cent rh to fully reach equilibrium, moisture content 24 per cent.

Hints on accuracy

Flooring substrates

Humidity sensors must be calibrated before use; this is usually carried out before arriving on site. **Mechanical hygrometers** require a covering of bubble wrap to keep them at a moderate humidity and temperature. During the journey they should be kept out of the sun. A secure place in the boot of the vehicle, free from mechanical vibration and damage, is recommended. Paper and natural hair hygrometers tend to wander in their calibration during travelling. Synthetic fibre hygrometers are relatively stable.

All artificial heating and dehumidifiers should be turned off at least **four days** before the test. This is to ensure that a false reading is not achieved. Air conditioning counts as an artificial aid, and must be turned off.

An "equilibrium" figure must be reached before the test is considered as complete. This is when at least two consecutive readings show no change at four- or 24-hourly intervals as appropriate to the form of construction. For example, recorded consecutive values of 65, 67, 68, 69, 69, 69 show that an equilibrium value of 69 per cent rh has been reached and the substrate is safe to receive floor covering if the maximum permitted value is 75 per cent rh.

Effective sealing of the insulated box to the substrate is essential. Any leakage may give low results and hence a false indication of readiness for laying the flooring. Insulated boxes should be regularly inspected to ensure that seals are functioning correctly. Protection against disturbance from other site operations is essential.

When the floor surface has previously been covered with impervious sheet to reduce the time required for the instrument to be in position on the floor ensure that a "false" equilibrium is not achieved. The BS suggests that equilibrium may be achieved with in two to four hours of positioning the instrument. This may be the case for isolated screeds, but for thicker constructions it is less likely. In both cases it is strongly recommended that the instrument should be left overnight and then readings taken at four- or 24-hourly intervals as appropriate to the form of construction.

Record ambient temperature and rh and check that adequate ventilation has been provided to remove moisture from the building.

Beware of just taking just one reading to represent large areas of flooring. It is recommended to place one humidity sensor to 75 m^2, or to move the equipment around having completed the test on one area.

Mechanical hygrometers are known to "drift" in calibration, therefore, it is wise to check the calibration before and after each test (location) and apply a correction for the "drift" ie assume a linear correction over the time period between checks.

When using humidity sensors for other applications:

- ensure that the sensor is always in equilibrium temperature with the material being measured and that the seal is airtight. Temperature variations during comparative testing should be avoided or minimised
- always brush out and vacuum clean drilled holes before inserting the sensor to avoid 'unknown' powdered material contaminating it and/or the surface of the hole
- check the battery level before taking meter readings. It is recommended that they be changed at 20 per cent potential drop; some meters are provided with a battery-warning indicator
- record both temperature and rh data from the sensor location. Also record ambient temperature and rh and correlate to sensor readings
- humidity sensors will require periodic recalibration. Calibration is easily achieved if the test arrangement requires the introduction of the sensor into the air pocket to obtain a reading. This technique requires care to avoid undue disturbance of the air and an adequate time for the sensor to reach equilibrium. Otherwise by careful design "built-in" sensors can be replaced at regular intervals with recalibrated sensors: again allowing time for the "new" sensor to reach equilibrium
- when inserting **electronic rh probes** into prepared sealed air-pockets ensure they are in place for equal time periods, eg 30 minutes per hole (minimum). Note: the time period above is directly related to the time the probe is exposed to the relevant saturated salt solution when being calibrated. (See manufacturer's recommendations for calibration)
- when using **wooden plugs**, fit a plastic strip to stop the plug from contacting the sides of the hole thus preventing salts being absorbed by the wood, eg from concrete. On removing the plug check for staining or signs of salt contamination, eg abnormally high reading when using an electrical resistance meter.

5.5.2 PRINCIPLES

Property

Flooring substrates

The humidity sensor measures the relative humidity of an enclosed and sealed air pocket above the damp substrate. Once the air and the substrate have come to equilibrium, the rh of the air is considered equivalent to that in the substrate.

The test method was developed when the Building Research Establishment was asked to look into the problem of moisture affecting the lignin paste adhesive when used in conjunction with linoleum in 1947. BRE, through various tests, found that the moisture level from rising water vapour through a cementitious screed could be measured with a hair hygrometer, and that the figure for the breakdown of the lignin paste was 80 per cent rh. To err on the safe side, and taking into account the test equipment accuracy, the safe limit adopted for laying flooring coverings was set at 75 per cent rh.

Mechanical hygrometers have a sensing element, which, as the result of small changes in length with change in humidity, is geared to move a needle on a clock face, see Figure 5.5.1 (left).

Electronic rh probes have a humidity-sensing film sensor that produces an electrical change, eg in capacitance, which can be measured using an appropriate meter, see Figure 5.5.1 (right).

Wooden plugs are hygroscopic, the moisture level in a plug, after a few weeks, can be simply read using an electrical resistance meter, or directly measured by oven drying.

Figure 5.5.1 *Showing a mechanical hygrometer (left) and an electronic rh probe (right)*

Test basis

Flooring substrates

There are many models of hygrometer boxes that are used to monitor the rh of an air space formed on the material surface; some incorporate an aperture to insert an electronic rh probe. The basic British Standard model is illustrated in Figure 5.5.2.

Mechanical hygrometers and **electronic rh probes** are calibrated to directly measure the rh of the air space in which they are placed.

A **wooden plug** sealed into a cavity absorbs moisture to reach equilibrium with the surrounding material.

Measurements

Humidity sensors are usually calibrated to directly measure rh, which is the ratio of the actual vapour pressure to the saturation vapour over a liquid water surface at the same temperature. Values are commonly expressed in whole units of a percent and written as X per cent rh.

For actual vapour pressure, e, and saturation vapour pressure e_s:

relative humidity (per cent) = $(e/e_s) \times 100$

The equilibrium relative humidity (ERH) (over a substance) is the value of relative humidity of the air at which there is no net exchange of moisture with any nearby substance. This is used for indirectly indicating or controlling the condition of moisture sensitive materials. Material equilibrium rh data can be related to moisture content, particularly for wood where conversion tables are available for various wood species.

Units and scale

The scale is 0–100 per cent.

Note: some instruments may display values in excess of 100 per cent rh. This generally indicates that condensation has formed, ie in wet conditions where the relative humidity is greater than 95 per cent, or that the sensor system is malfunctioning.

Sensitivity

Most instruments can be read to 1 per cent rh. The measured rh, as indicated, can be considered independent of temperature in the range 5–30°C, however actual rh is sensitive to temperature change.

Accuracy

Mechanical hygrometers will give a reading accurate to ±5 per cent rh if the equipment is used correctly.

Electronic rh probes will generally have an accuracy of ±2 per cent rh below 75–90 per cent rh, depending on the sensor type, decreasing to ±3 per cent at 100 per cent rh.

Wooden plugs will generally have an accuracy of ±3 per cent rh.

5.5.3 EQUIPMENT

Components

Flooring substrates

An instrument to register relative humidity, eg mechanical hygrometer or electronic rh probe and meter.

An insulating and isolating box that complies with BS requirements to obtain an excellent vapour seal and good thermal insulation around the instrument. Boxes are normally fabricated from polystyrene, wood packed with polystyrene or plastic also insulated with polystyrene. The internal faces of the box must act as a vapour barrier. Any window in the apparatus, to allow viewing a mechanical hygrometer, must be of glass or acrylic sheet. Note: Purpose-made insulated boxes are available commercially.

Preformed butyl sealant tape or other suitable material to seal the box to the substrate.

Rubber mat or polyethylene sheet at least 1 m^2 to simulate an impervious covering and adhesive tape to seal the edges of the mat or sheet to the substrate.

When using **electronic rh probes** for other applications:

- drill and, typically, 20 mm drill bit (both selected for material to be drilled and the diameter of the probe)
- small nozzle vacuum cleaner
- plastic pipe to fit into hole with end cut at an angle
- small piece plastic sheet
- hand-held humidity and temperature meter/thermometer.

Note: some suppliers of probes provide purpose made sleeves for measurement in materials, specifically concrete.

When using **wooden plugs**:

- ramin dowel
- plastic strip and rubber band
- rubber bung to seal hole
- closed cell foam or polythene or cling film to fill between the plug and the bung
- drill and, typically, 14 mm drill bit (both selected for material to be drilled and the diameter of the plug)
- hand-held electrical resistance meter (for timber) and temperature meter/ thermometer.

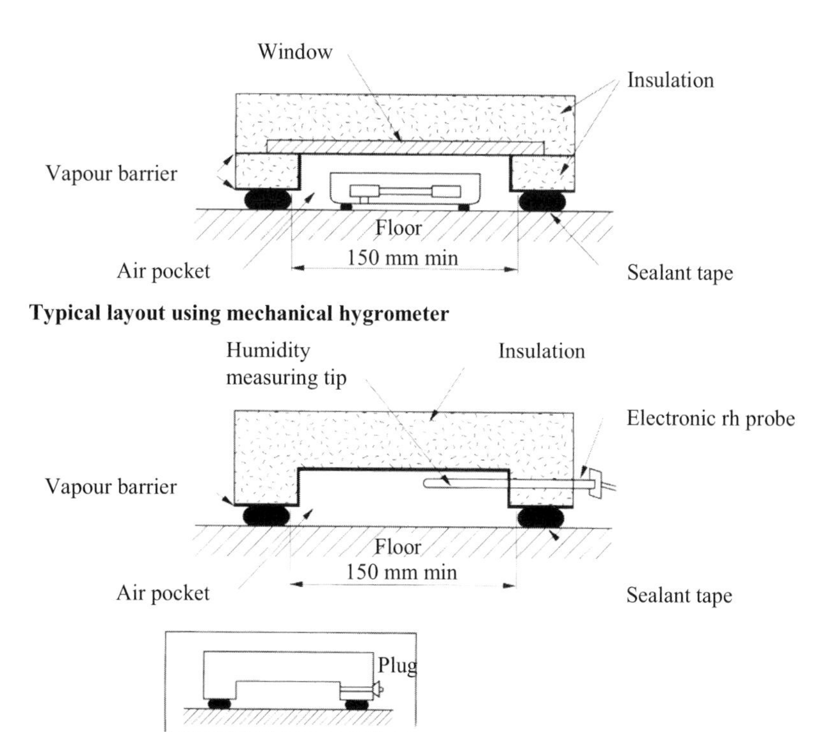

Figure 5.5.2 *BS Codes of Practice test method (courtesy of British Standards Institution)*

Size

Flooring substrates

The size of the box can vary but the diameter of the pocket of air trapped must be at least 150 mm.

For manual operation using **electronic rh probes** the meter will be hand held with a probe typically 360 mm long × 10–12 mm diameter. Smaller probes, down to 4 mm in diameter, are available.

Wooden plugs will typically be 50 mm long and 8–10 mm in diameter.

Weight

Flooring substrates

The weight of the box and humidity sensor varies with size but an average weight is approximately 1.5 kg.

An **electronic rh probe** and meter typically weigh 400–700 g. A **wooden plug** and meter will have a similar weight.

Storage

It is recommended that all instruments are stored in the manufacturer's cases. It is important that instruments and sensors are stored in a stable dry environment.

Calibration

Flooring substrates

For **mechanical hygrometers** an airtight sealed container with a tray or rods is required to prevent it from falling in the liquid: purpose designed calibration tanks are generally available for **electronic rh probes**. It is recommended to check the calibration of **mechanical hygrometers** before and after each test and apply a linear correction for any "drift".

The **humidity sensors** are placed inside the sealed container over a saturated sodium chloride solution, which gives 75.5 per cent rh at 20°C, and allowed to come to equilibrium. In the event that adjustment is required, **mechanical hygrometers** will have to be removed from the receptacle. After adjustment, they should be replaced and the calibration carried out again. It may be necessary to repeat this procedure several times until an accurate reading is achieved. Particular care is needed in controlling the purity of the salt solution, sealing the humidity sensor into the test container and in establishing a controlled/stabilised temperature during the test.

Electronic rh probes can usually be adjusted without the need to disturb the sensor. However they have two adjustments, span and offset, hence if it is necessary make any adjustment there is a need to verify the reading at a different rh values, eg lithium chloride (LiCl) 12.4 per cent rh and potassium sulfate (K_2SO_4) 97.2 per cent rh at 20°C. Again the process should be repeated, allowing time for the probe to come to equilibrium, until no further adjustment is required.

The recommended calibration interval for **electronic rh probes** is 12 months however if they are subjected to extremely high humidity, eg above 90 per cent for long periods the frequency should be reduced to, typically every six months (seek the manufacturer's advice).

Wooden plugs are cut from new ramin dowel or a similar standard wood for which calibration, ie table of moisture content against ambient rh, is available. Where additional accuracy is required the sensors should be calibrated over salt solutions.

Figure 5.5.3 illustrates the potential changes in "calibration" when relating ambient relative humidity to moisture content if the material is subjected to wide variations in moisture condition. Plugs that have absorbed salts from concrete or have been oven dried should not be re-used. When using an electrical resistance meter check the instrument reading against the standard resistance, see Section 5.4.3.

5.5.4 METHOD OF OPERATION

Requirements

Samples

Flooring substrates

It is recommended to place one humidity sensor to 75 m^2. Find the wettest parts by inspection or, if appropriate, using an electrical capacitance or resistance meter, and place the humidity sensor in this area. Note: cementitious substrates tend to dry from the centre towards the perimeter of the area).

For other applications electrical capacitance or resistance meters can be used to assist in the location of appropriate positions to install humidity sensors.

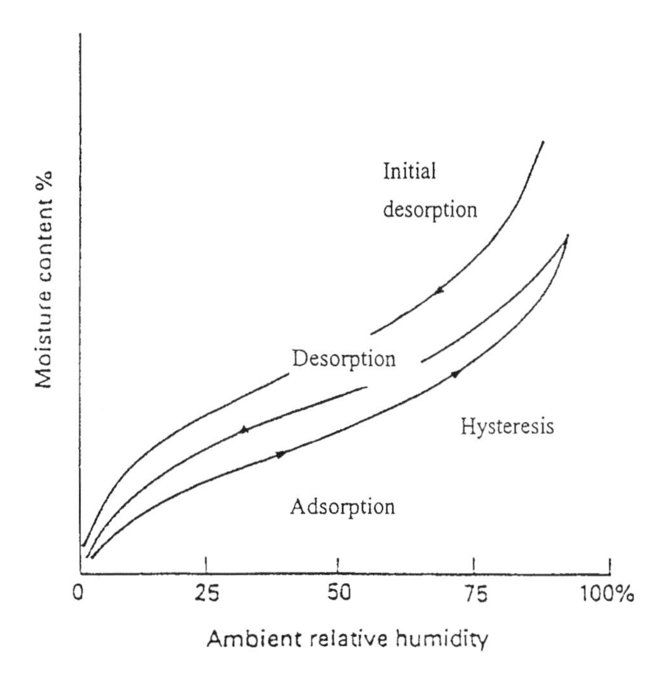

Note: the connection between moisture content and relative humidity at equilibrium at a constant temperature is called the sorption isotherm or simply the isotherm.

Figure 5.5.3 *Typical adsorption and desorption isotherms showing hystersis in a typical building material*

Environmental

In the short-term, humidity sensor data is not sensitive to general weather conditions as it is recommended that the humidity sensor be provided with both a vapour seal and good thermal insulation. However, the weather (minimum measurement requirement of temperature and rh) and surface conditions in the area of study should normally be recorded as part of the evaluation. Temperature measurement at the gauge location may additionally be required for temperature correction, eg on the meter reading and/or conversion to vapour pressure.

Situational

Flooring substrates

An area of at least 1 m^2 is required to carry out testing in accordance with the BS test method.

For other applications the location for humidity sensors can be at any point of interest provided that the effects on the building operation, any finishes and moisture and vapour migration are considered. It is important to avoid water running past the seal as this increases the risk of water penetrating the enclosed air space. This might be a particular problem where the sensor is inserted down into a hole in a floor slab or screed; on vertical surfaces minimise the risk by inclining the hole to make it self-draining.

Human resources/ skills

The user needs to be familiar with the use of the instrument, have an understanding of or access to information relating to acceptable levels of humidity within materials and understand the importance of having a stable temperature environment

Initial trial testing

Before installing humidity sensors and using them on site the staff involved should make up the basic sets of the required equipment and check that the materials and procedures to be employed have no significant effect on the sensor calibration.

Use

Flooring substrates

Mechanical hygrometers should be treated carefully. Dropping a mechanical hygrometer can cause serious problems, especially if it is glass enclosed. They are delicate, should be handled carefully and calibrated regularly; rough handling will cause drift.

Make sure that all-artificial heat, de-humidifiers and air conditioning has been turned off for at least four days before testing.

Having found a suitable position for the test, place the humidity sensor and insulating box in the required test position and use a preformed butyl sealant tape, or similar, to hold the box in position and seal it to the test substrate. Do not use silicone sealant or other material that could affect the test result. Effective sealing of the insulated box to the substrate is essential. Any leakage may give low results and hence a false indication of readiness for laying the flooring. Insulated boxes should be regularly inspected to ensure that seals are functioning correctly.

Protect the equipment by some form of barrier and warning signs so that people will not trip over it or disturb it.

Adequate protection against disturbance from other site operations is essential.

Take and record the first reading after 4 hours in the case of isolated screed substrates. For thicker constructions of direct finished concrete or bonded screeds, the first reading should be taken after a period of at least 72 hours. Refer to the appropriate BS Code of Practice or manufacturers recommendations for the acceptable maximum value of rh. The most usual maximum value is 75 per cent rh. If the first reading is 5 per cent above the specified maximum rh, remove the equipment and return at a later date.

Take and record further readings at four-hourly intervals, or 24-hourly intervals in the case of thicker constructions, until a genuine equilibrium is reached.

To minimise the time required for the instrument to be in position on the floor, test areas can be covered with 1 m^2 impervious mats or polythene sheets sealed at the edges. These are left in position for three days for isolated screeds or seven days for thick constructions. The instrument is then sealed to the centre of the covered surface, after uncovering the required minimum area. The BS indicates that equilibrium will then be achieved within two hours, but recommends taking further readings the following day for confirmation.

Note: thicker constructions, such as power floated concrete or screeds bonded onto concrete slabs can take several weeks to reach equilibrium.

Although the BS defines equilibrium as two consecutive readings showing no change, more consecutive readings give greater confidence. Watch out for two or three results being equal and the fourth reading rising. In this case the substrate is letting the moisture out intermittently. Wait until the reading drops or is stationary for several days.

The 1987 BS Codes of Practice put the responsibility on all parties as to the condition of the substrate, meaning if the one party has the test carried out and reports that it is satisfactory, the other parties also have a duty of care to check/witness the test. The main duty of care, however, must always remain with the party responsible for carrying out the test. It should faithfully record and pass all relevant information in writing to the party commissioning the tests. The **1996 revision** is less ambiguous. It strengthens the statement that these responsibilities are defined in preliminary contract discussions and recommends that they are taken by the main contractor.

When sealing an **electronic rh probe** or **wooden plug** within a porous material:

- verify that it is safe to drill into the material at the selected location; ie check for electrical cables, pipes and reinforcement
- drill a hole of the appropriate diameter for the humidity sensor to the required depth; where appropriate the hole should be inclined to be self-draining
- brush and vacuum the hole to remove all powered material
- when available, line the inside of the hole with a plastic sleeve to the required depth of the humidity sensor
- insert the humidity sensor to the required depth, a rubber O-ring can be used to provide a seal within the hole and a closed cell foam plug inserted to fill the remaining air space providing insulation (see Figure 5.5.4)
- if required when using an **electronic rh probe**, the hole can be sealed for at least 48 hours, using a rubber plug, prior to inserting the sensor (see Figure 5.5.5)
- insert a rubber plug, or similar, at the surface and provide protective cover to ensure that the seal remains undisturbed
- take and record the first reading after **30 minutes** in the case of a previously sealed hole and for other applications wait an appropriate time for sensor and air space to stabilise, minimum of 48 hours
- take and record further readings as required depending on the construction thickness and purpose of the measurements.

When investigating long-term performance of a building or repair, the determination of in-situ moisture and humidity gradients provides essential data. The first step is a check for water or water vapour availability and temperature and vapour gradients within the material for the normal cycles of exposure (day and night, summer and winter, rain and sun). The disposition of the cavities for the humidity sensors must be related to the expected gradients within the construction and the frequency of reading must be related to the cycles of wetting and drying to be expected, particularly if rh readings are to be recorded manually.

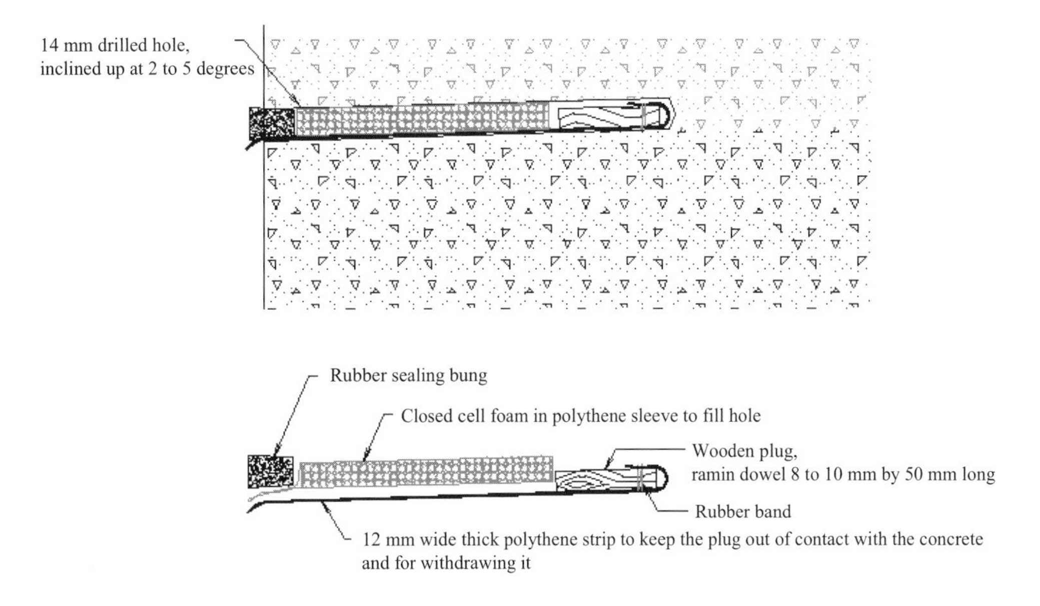

Figure 5.5.4 *Showing the installation of a wooden plug into a wall*

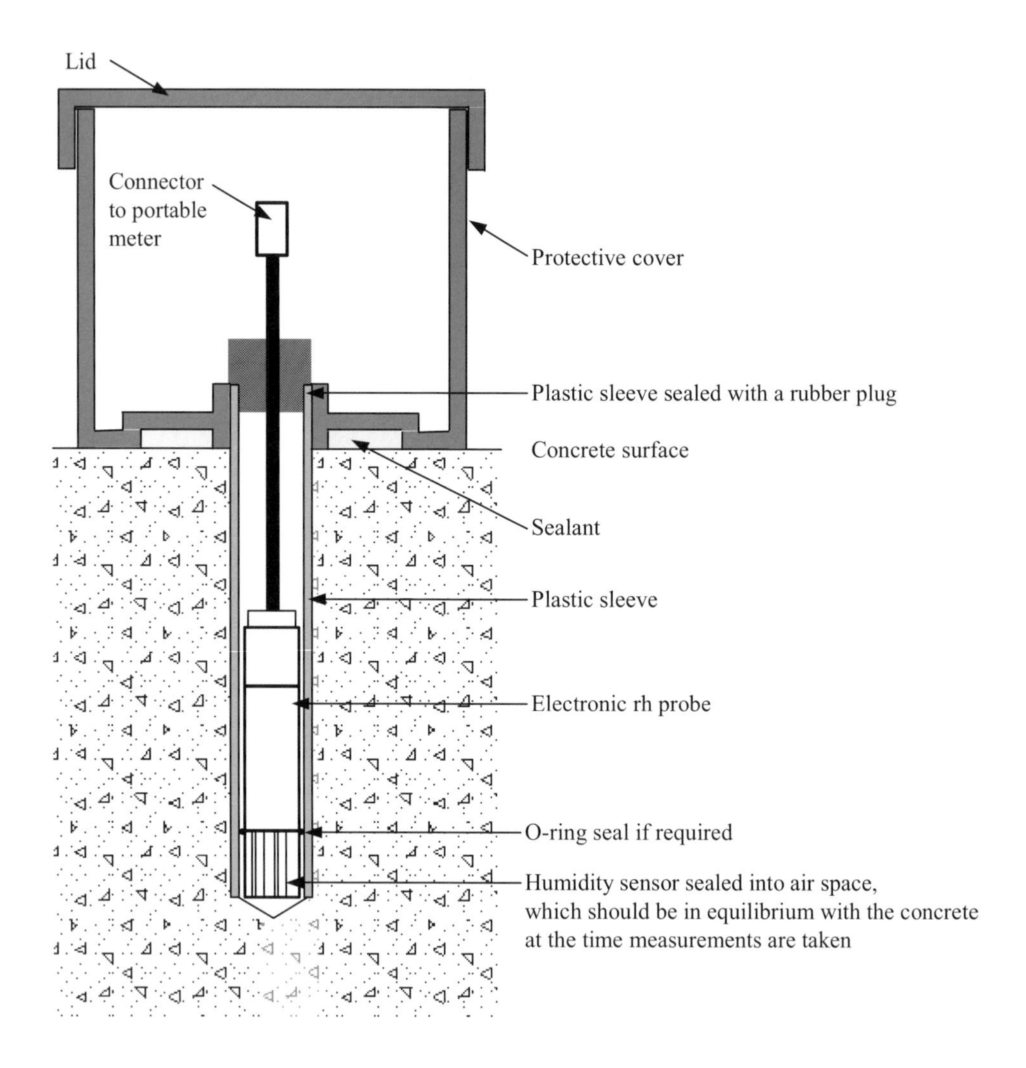

Figure 5.5.5 *Showing the installation of an electronic rh probe in a concrete floor*

Data recording

Flooring substrates

All readings should be recorded showing date, time, temperature and position of the equipment.

A similar recording procedure should be used for manual readings taken from **electronic rh probes** and **wooden plugs**. This data can be input and stored, after a number of readings have been taken, onto a PC and using the relevant software, drying profiles can then be produced. Automated long-term monitoring systems should record the data for direct transfer to a PC for collation, analysis and presentation of the information (see Section 6 *Automated long-term monitoring*).

Results

Flooring substrates

If the initial reading is higher than, say, 85 per cent rh, remove the equipment and re-test the floor after allowing an appropriate time for the substrate to dry.

The readings taken from a humidity sensor indicates the equilibrium rh of the material and is independent of the type and quality of the material. The data is generally used to demonstrate that no excessive moisture remains in the building material. When interpreting the data it should be remembered that, for a given material type, moisture contents of 3 per cent by dry weight could mean humidity of 40–95 per cent rh, depending on the composition of the material and whether the material was being moistened or dried. Conversely a given equilibrium relative humidity may equate to a wide range of moisture content, for example 1.5–4 per cent. Hence, in addition to obtaining rh data it is necessary to understand the material properties and have knowledge of the history of the operational conditions for that element.

It is generally accepted that equilibrium relative humidity readings of 75 per cent in a material signify that is in safe state to receive a finish. It is also important that where a dampness problem exists, rh and temperature measurements at different depths should be taken through the construction thickness at selected locations. This data will provide moisture profiles and can be converted to give the vapour pressure gradients that can be used as aids in the diagnosis of the problem.

Specification

Flooring substrates

"Carry out tests for measuring the dryness of the substrate using a BS Code of Practice hygrometer test". The appropriate BS should be specified.

Section 3.4 *Standards and specifications* gives further general guidance.

5.5.5 CASE STUDY

Flooring substrate

Construction A leading retail chain store had recently been modernised with new vinyl tiles on a flooring substrate that was a conventional screed on concrete slab, some of which had been replaced.

Reason Because some of the new vinyl tiles were lifting, humidity readings were to be taken of the flooring substrate.

Results The rh measurements were to be carried out using an electronic rh probe and meter. Holes were drilled at various locations through out the store at discreet points. These were then vacuumed out and fitted with a plastic tube cut at an angle at the end and then sealed at the top and left for two days. The probe was left for half an hour in the area where the first measurement was required to allow the instrument to reach equilibrium with the conditions in the store. The probe was then carefully inserted into the prepared holes, allowed to stabilise and rh readings taken. The readings showed that the rh was 97–99 per cent within the flooring substrate.

Success The test results indicated a high probability that the moisture levels within the substrate were too high prior to the new flooring being laid.

Assessment The BS 8203 Code of Practice requirements for the application of floor coverings were unlikely to have been adhered to, 75 per cent rh being the accepted level for surface rh measurements. The measurement within the material may be slightly higher than at the surface but not significantly higher.

5.5.6 RELEVANT DOCUMENTS

Standards and legislation

British Standards Institution, *Code of practice for installation of resilient floor coverings*, BSI, London, BS 8203: 1996

British Standards Institution, *Code of practice for flooring of timber, timber products and wood based panel products*, BSI, London, BS 8201: 1987, (ISO 631 NEQ; ISO 1072 NEQ)

British Standards Institution, *Code of practice for installation of textile floor coverings*, BSI, London, BS 5325: 1996

Guidance

None

Key reading

Nilson, L, *Hygroscopic moisture in concrete – drying, measurements and related material properties*, University of Lund, Gothenburg, Sweden, Report TVBM-1003, 1980

National Physical Laboratory and The Institute of Measurement and Control, *Guide to the measurement of humidity*, InstMC, London, 1996

Structural Studies & Design: *Notes on application of wooden plugs for the determination of insitu rh and moisture in concrete and other porous materials*, trade literature

Wood, J G M, *Methods for the control of active corrosion in concrete*, 1st International Conference Deterioration and repair of reinforced concrete in the Arabian Gulf, Bahrain Society of Engineers and CIRIA, October 1985

Wood, J G M, Durability design: form, detailing and materials, *Building the future*, pp 23–32, Garas, F K (ed), E & FN Spon, London, 1994

Wood, J G M, Nixon, P J and Livesey, P, *Relating ASR structural damage to concrete composition and environment*, pp 450–457, A Shayan (ed), Proceedings 10th International Conference Alkali-aggregate reaction in concrete, Melbourne, 1996

Parrott, L J, *Factors influencing relative humidity in concrete,* Magazine of Concrete Research Vol 43, No 154, pp 45–52, March 1991

Parrott, L J, *Moisture profiles in drying concrete*, Advances in Cement Research Vol 1, No 3, pp 164–170, July 1988

Jensen, V, *Use of wooden plugs to determine relative humidity in concrete*, in Norwegian. English paper in preparation for 11th International Conference Alkali-aggregate reaction in concrete, Quebec, June 2000

5.6 Microwave moisture meter

5.6.1 APPLICATION AND USE

The hand held microwave moisture meter described in this section is a new piece of equipment, produced by one manufacturer, and has not been fully tested or proven in use.

Principle

The microwave moisture meter is used to evaluate the "free" moisture content (capillary moisture plus hygroscopic moisture) by measuring the dielectric constant of the material being tested. It works on a similar principle to a capacitance meter, but at a much higher frequency that makes it less susceptible to impurities such as salts. By simultaneously measuring the amplitude as well as the frequency of the electronic signal, it is theoretically possible to make the readings independent of the density of the material.

Other names

Higher frequency power absorbtion method.

Application

A hand held microwave moisture meter is a recent development based on microwave moisture sensors used in the production of concrete in large industrial mixers as well as the general moisture measurement of aggregates (sand and gravel). Although the advantages in the use of microwave absorption for the measurement of moisture are widely recognised, until recently the size, weight, and power requirements of a practical system have made this method of measurement possible only in fixed installations. The small, battery operated hand held meter, which can be used for in-situ measurement, is a significant development in this respect. It is non-destructive, safe and convenient to use, and is claimed to be highly accurate.

Usage

The meter incorporates full density correction.

Materials

The meter was designed primarily for the measurement of moisture in concrete blocks and pipes. It can be used for any building material that contains "free" water.

Elements

Mainly walls and floors.

Use with

The meter measures moisture content directly. It can also be used to indicate damp areas in a wall or floor so that samples can be obtained by drilling and used for chemical analysis and/or other forms of moisture determinations eg oven drying, calcium carbide method.

Health and safety

The total amount of power radiated to provide a measurement of moisture is approximately 5 milliwatts. This is low enough not to present any health hazard. It is very small compared with the 750–1000 milliwatts radiated from a hand held telephone, or to the leakage often occurring from a microwave cooker. When in use, nearly all the radiation is absorbed by the material under test.

Costs and time

The microwave moisture meter gives instant measurements of moisture content. However, being a new and unproven device, time and costs may need to be allowed for familiarisation and comparative testing. The initial cost of instrument is much higher than electrical resistance or capicitance meters.

Relevance

Theoretically a microwave device provides a measure of the "free" water that is independent of the density of the material. This meter may therefore prove to be the best indicator of moisture when investigating damp related defects. It may also prove particularly useful in the situations when other techniques give "false" readings due to the presence of salts (see Section 3.3 *Moisture measurements*). **However the microwave meter described is innovative and has yet to be proven.**

Advantages

- There is no direct electrical contact with the material being tested.
- No change in calibration is needed for samples of different density; by measuring two parameters, which are both dependent on density and computing the results, the readings are independent of density. There is therefore no requirement for conversion tables or charts.
- Microwave absorption is affected by fewer substances, which have the same affect as water and cause erroneous readings, than other moisture measurement methods. In particular it is not significantly affected by any salt content.
- The system basically measures only free water and has very little response to bound water, absorbed water, water of crystallisation, chemically absorbed water, etc. This means it responds primarily to "unwanted" water.
- Can be used for powders and solids.
- Light, small and portable; very easy to use.
- Good repeatability of measurements.

Prompts and pitfalls

- Although the meter will work on most porous materials, errors may occur with certain ceramics.
- The presence of metals can cause false readings due to reflections from metallic surfaces and standing waves from metal parts. For example, care must be taken when measurements are made in reinforced concrete as steel near to the surface can cause errors.
- The effect of water on the microwave absorption is much less at the edge of the field than near to the body of the instrument, due to the fact that the intensity of the electromagnetic field decreases inversely proportional to the square of the distance from the source. Water near the surface of the material will thus have the greatest effect on the readings. Any surface condensation will result in high readings.

Hints on accuracy

- The surface of the material being tested should be as flat as possible.
- There should be no traces of surface condensation.
- The sensitive surface of the meter should be placed firmly on the test material.

5.6.2 PRINCIPLES

Property

The unit, shown in Figure 5.6.1, incorporates a microwave generator and resonator that produce an electromagnetic field external to the end of the instrument case. This field is allowed to interact with the material to be assessed by placing the instrument in contact with the material. Microwave energy is absorbed by water therefore by measuring the absorption the amount of water present can be determined. The penetration of this field is approximately 50 mm.

Test basis The percentage moisture is inversely proportional to the density of the material containing it; however, in many measurement systems the density of the material is taken as constant leading to severely limited application and accuracy. The microwave meter avoids this drawback by measuring two parameters separately and simultaneously: the energy absorbed from the microwave field by the water and the phase change caused to the field of the resonator. Both of these parameters can be expressed mathematically and both are dependent upon the density of the material. By solving these mathematical expressions density can be eliminated and the percentage moisture calculated. This computation is done by an in-built microprocessor and thus the readings displayed are independent of density.

To obtain the maximum accuracy and stability in a sensor, systems are incorporated to ensure that the amplitude and frequency of the resonator do not change when measurements are being made. By providing negative feedback to the resonator such that as the electrical damping, due to the energy absorbed by the moisture, is increased, the output from the generator is increased to keep the amplitude constant. Similarly as the resonator frequency changes due to the moisture, a feedback system controlling the voltage on a varactor diode is altered such that the resonator frequency is maintained constant. From the amount of feedback required for this operation the components of the permittivity can be calculated, and translated into moisture content.

Measurements The microwave meter is normally calibrated to measure wet weight moisture content directly. It can also be calibrated for dry weight moisture content if required. It should be clear which basis is being used, as the results are very different (see Section 3.3 *Moisture measurements*).

Units and scale Percentage moisture content in the range 0–40 per cent by wet weight on a digital scale.

Sensitivity The scale has a resolution of 0.01 per cent.

Accuracy Estimated as ±0.1–0.5 per cent for aggregates and sand based on the accuracy of the fixed moisture meter installations, but this has not yet been proven for the hand held meter.

5.6.3 EQUIPMENT

Components The instrumentation is housed in a hard case that is internally screened by copper covering the whole interior of the case except for the microwave window at the end of the case, and for the display window on the rear.

The internal batteries are nickel metal hydride and are capable of providing six hours of continuous operation. They are charged by means of an auxiliary charger provided.

Size Hand held.

Weight 200 g.

Storage The meter should be stored in its carrying case in a dry environment.

Calibration Calibration is carried out on manufacture and an annual verification is recommended.

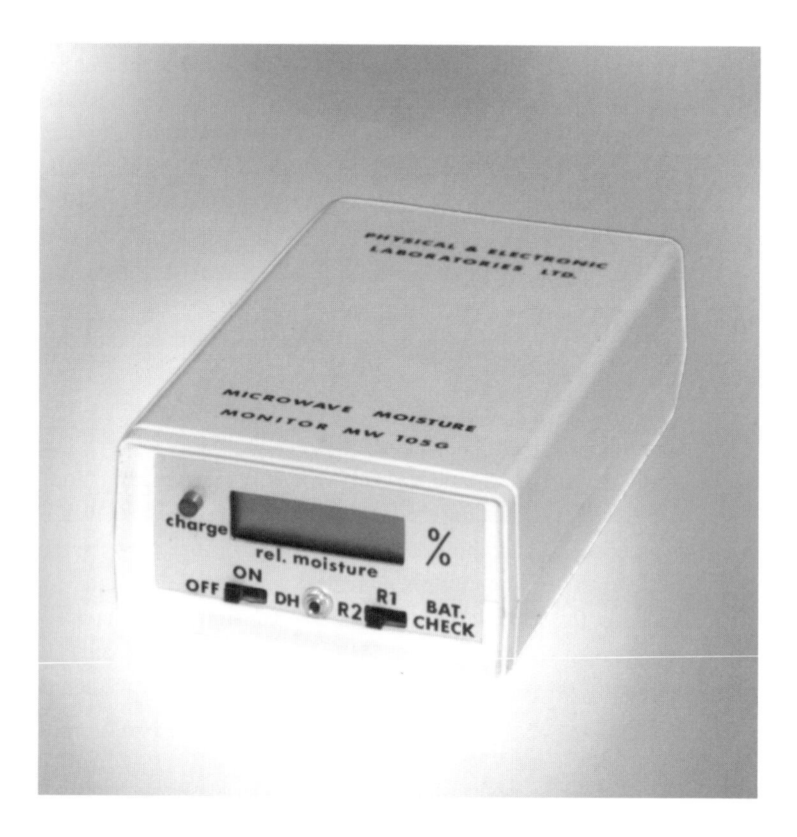

Figure 5.6.1 *Microwave moisture meter*

5.6.4 METHOD OF OPERATION

Requirements

Samples Measurements should be taken at appropriate intervals, depending on the objectives and element or material being tested.

Environmental The equipment must be kept dry at all times. The presence of surface dampness (eg condensation, dew, rain) is likely to yield "high" readings.

Situational As with all hand held instruments it may be difficult to read when used in awkward locations; a data hold function is provided to assist the operator.

Human resources/skills A high level of expertise is not required for using the equipment. However, some basic training and knowledge, and involvement with comparative testing for validation may be necessary. Care must be taken in interpretation and presentation of the data.

Initial trial testing Until there is sufficient experience using this innovative equipment comparative trials will be necessary to verify actual performance. It is not envisaged that initial trial testing will be necessary once the equipment has a proven track record.

Use

Moisture content readings are taken by placing the sensitive end of the case against the sample to be tested. Firm but not excessive pressure is applied and to maintain accuracy a flat section of the sample should be chosen.

Data recording

Wet weight moisture is indicated directly on a digital display. Two ranges may be provided for different moisture levels. There is a display hold facility to lock the display reading for 20 seconds when measurements are made in dark or inaccessible places.

Results The results, directly read from the meter, should be recorded and checked for consistency.

Specification There is no standard specification. See Section 3.4 *Standards and specification* for general guidance.

5.6.5 CASE STUDY

The meter is commercially available but has not been fully tested or proven in use.

5.6.6 RELEVANT DOCUMENTS

Standards and legislation None

Guidance None

Key reading Assenheim, J G, *Moisture measurement in the concrete industry,* Concrete Plant and Production, Sept/Oct 1993

5.7 Nuclear moisture gauge

5.7.1 APPLICATION AND USE

Principle

The hydrogen atoms in a material, principally in the moisture, are detected using a radioactive source and counted.

Other names

Roof reader, roof moisture gauge, neutron probe moisture meter.

Application

The principal application for the use of portable nuclear moisture gauges in buildings is in their use for roof surveys to assess moisture trends. They may, however, be used on any flat surface to plot trends of moisture within a structure but due to their relatively high weight (up to 25 kg) they are not really very suitable for walls or vertical surfaces.

They are widely found in use on construction sites in the form of a combined gauge used for determining moisture and density properties of compacted soil, concrete or bituminous materials.

The moisture component of these combined gauges is addressed in this section; indeed some gauges are manufactured without the density component and are specifically for looking at roof moisture.

The basis of this test is not normally one of determining actual moisture content but is a means of observing moisture trends. Testing is carried out over a grid to enable a map of moisture to be made for a roof such that any faults may be easily highlighted and remedial work may be targeted in only the areas where it is necessary. Some gauges have displays that will show actual moisture content. However careful calibration is necessary to enable these to function with any degree of accuracy and is reliant on the density of the material being accurately measured.

They have the advantage of being able to measure from the surface down into the material to a theoretical maximum depth of 280 mm on materials with no moisture or hydrogen present in them.

Usage

Materials

The meter may be used on a wide range of materials but has the limitation that the chemical composition of those materials needs to be consistent so that effective comparative measurements may be made.

Elements

The meter is mainly suited to measuring moisture in roofs and floors but it could be possible to adapt it to measure other elements, such as walls, providing some kind of mounting rig was used and providing the base of the gauge may be held flat against the surface.

Use with

It is not usually necessary to use any other test methods with the portable nuclear moisture gauge. However, to ensure that results are valid, a careful visual recording of material under test will enable any inconsistencies in results to be explained by material type.

Health and safety

These devices use small radioactive sources that are governed by regulations for ownership use, transportation, etc in the UK. The specific applicable regulations are:

- The Radioactive Substances Act 1993
- The Ionising Radiation Regulations 1985 (due to be updated in January 2000)
- The Radioactive Material (Road Transport) (Great Britain) Regulations, 1996.

The users of this equipment must satisfy all aspects of these regulations and the manufacturer's instructions. The radioactive dose received by operators will always stay very low, which can be demonstrated by the use of personal radiation badges. Many construction-testing facilities already have all of the necessary mechanisms in place to allow them to use their instruments on the testing of compaction of a range of construction materials. Section 3.5 *Health and safety* gives further general guidance.

Costs and time

Readings can be obtained in as little as 7.5 seconds and therefore a large number of results may be obtained in a day. A grid pattern is recommended to obtain good plots of data and preparation of this may take longer than the actual tests using the meter. Obstructions across a surface will also cause a reduction in overall test time.

Another consideration in budgeting for a test is the cost of the health and safety issues that will require set up and running costs over and above those of non-radioactive equipment.

Relevance

The portable nuclear moisture gauge is particularly relevant to plotting moisture trends across flat roofs where it is necessary to determine which part of a roof may require remedial action.

Advantages

- An in-situ non-destructive test method that is quick and easy to use.
- Its particular advantage over other electrical meters is that it can measure moisture to a depth within the material and does not only give moisture readings at the surface.
- Proven technology of neutron thermalisation that can be used in most environmental conditions.
- Equipment widely in use for other tests eg compaction control of paving materials.

Prompts and pitfalls

- The method of relating hydrogen in a material directly to the water can have its drawbacks, as it is possible that hydrogen may be found within materials in other forms other than moisture. This in itself is not a real problem providing the "other hydrogen" is consistently bound in the material across the entire surface such that the readings will all be influenced the same and therefore the moisture trends will not be affected.
- Roofs with many objects sticking up may prove difficult to monitor using this method: it is always necessary to compensate for the vertical face of an object in close proximity (within 1.5 m) with the side of the measuring gauge. This may cause higher quantities of thermalised neutrons to be measured thus apparently higher moisture readings may be recorded.
- Health and safety issues of using a radioactive source must be observed.

Hints on accuracy

- If absolute moisture densities or moisture content readings are required rather than trends of variations in moisture it will be necessary to field calibrate the gauge to the material on test.

5.7.2 PRINCIPLES

Property These instruments work on the principal of neutron moderation. Hydrogen atoms in the material being tested slow fast neutrons emitted by a radioactive source. Slowed neutrons are detected and counts displayed are proportional to moisture (hydrogen) content.

Test basis This method of test measures **all** moisture present including molecular water contained in some materials ie concrete, plaster etc but providing this is contained consistently throughout the material on test its effect may be accounted for. A schematic diagram of the apparatus is shown in Figure 5.7.1.

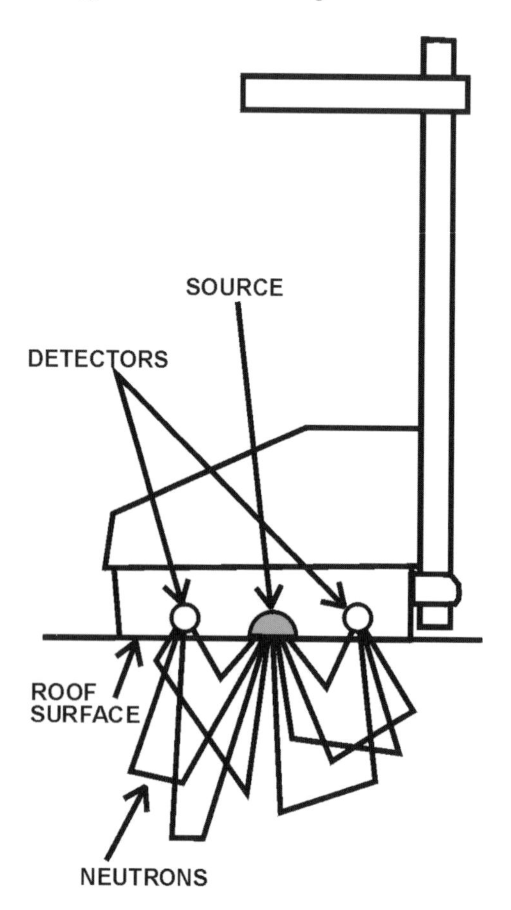

Figure 5.7.1 *Nuclear moisture gauge*

Measurements The display shows the basic radioactive counts that may be converted into units of moisture density, either externally or through internal software.

Units and scale The units of moisture density are kg/m^3 and the useful measurement range is 0–1000 kg/m^3.

Sensitivity The resolution is 7.6 kg/m^3 per radioactive count.

Accuracy Table 5.7.1 shows the precision from actual readings taken from known measurements. Each instrument will vary slightly dependant on its exact geometry and settings but the figures may be used as a reasonable estimation of all gauge precision.

Table 5.7.1 *An example of apparatus precision*

Moisture content	Precision kg/m³ Sample time		
kg/m³	7.5 s	15 s	60 s
0	4.74	3.35	1.67
100	8.46	5.98	2.99
200	10.98	7.77	3.88
300	13.03	9.21	4.61
400	14.79	10.46	5.23
500	16.37	11.57	5.79
600	17.81	12.59	6.30
700	19.13	13.53	6.77
800	20.38	14.41	7.20
900	21.55	15.24	7.62
1000	22.66	16.02	8.01

5.7.3 EQUIPMENT

Components The apparatus, shown in Figure 5.7.2, essentially consists of a portable gauge that has mounted in its base a small radioactive source usually of Americium 241 mixed with beryllium that emits a particular type of radiation known as neutrons. Also mounted in the base is a detector usually a Helium 3 Tube and circuitry to translate signals from this tube to a digital reading on a display panel.

The gauge is a single component and is supplied with internal rechargeable batteries that require the use of a separate charger.

On the combined density and moisture gauge units a reference block is also supplied which is used for accounting for source decay. However when using these gauges exclusively as a means for measuring moisture there is no real need to compensate for source decay. The source used has a half-life (time for the activity to be reduced by 50 per cent) of 458 years, ie in real terms there is no measurable source decay.

Separate items for marking out a grid pattern are recommended for use with the instrument.

Size The gauge is a portable device that will fit comfortably in the boot of any car.

Approximate dimensions of the transport container in which the unit has to be moved are 700 mm × 400 mm × 400 mm.

Weight Varies dependant on model used but is in the range 10–40 kg.

Storage

Due to the radioactive source the unit may only be moved in a TYPE A Yellow II label container that takes the form of the transport container supplied with the gauge.

A dedicated storage area has to be designated.

Calibration

Once factory calibrated there is no real need to recalibrate for moisture due to the long half-life of the source.

Figure 5.7.2 *Picture of roof moisture gauge and conventional moisture/density gauge*

5.7.4 METHOD OF OPERATION

Requirements

Samples A grid pattern is recommended to obtain good plots of data and allow an adequate normal distribution to be calculated (at least 100 readings are recommended).

Environmental These instruments use radiation detector tubes that require a high voltage circuitry to enable them to function. This high voltage is very moisture sensitive, and while the gauges are designed to be water resistant experience has shown that if they get wet problems with the counting mechanism may occur.

Situational All measurements are made in situ and can be influenced by vertical surfaces as outlined previously. The influence of these "side wall" effects depends on their construction but is not significant if they are more than 1.5 m away.

Human resources/ skills

No real skill is involved in using the instrument but all operators must be trained to satisfy Health and safety requirements. In addition some expertise is required in the interpretation of data and plotting of moisture profiles.

Initial trial testing

No trial testing is necessary unless direct moisture content readings are required when some calibration against core samples would be required.

Use

Mark out the grid pattern on the roof. A popular method is the use of two marked ropes placed parallel on opposite sides of the roof with a third rope stretched between them using spray paint to mark the grid intersections.

Make a scale drawing of the roof marking on the grid pattern and use this to carefully identify all roof structures, eg drains, heating and air conditioning units, ventilation shafts etc. If there are no structures that identify clearly roof orientation mark on plan north orientation.

Switch on and allow to meter stabilise (normally ten minutes).

Select time interval on display or through switch; 15-second count times have been found to be suitable, shorter count times give faster results but are not as precise.

Start taking readings on a grid pattern, recording data at each intersection.

Make a note against any reading where it is obvious there is a change in the roofing material tested.

Data recording

Data is best recorded as above on a grid pattern.

Results

Careful interpretation of the data is essential to obtain a good clear plot of moisture profile across the roof. Although it may be possible to use the data collected in a direct manner, it is far more useful to plot areas of extremely high, very high, high, normal and low (or otherwise categorised) moisture.

Construction of a frequency histogram (use of a spreadsheet helps) and the subsequent calculation of standard deviation and ultimately determination of wet and dry areas are necessary steps to enable the plotting of data to show lines of equal wetness. The main purpose behind plotting a histogram is to identify those areas that may be considered dry and eliminating them from the plot to leave just those areas that may be considered wet. Those are the ones of interest.

As radiation is statistical in its nature when looking at the total data recorded across the grid those readings taken within a limit of three standard deviations of the mean are considered as a normal dry material. There should be no readings lying below three standard deviations – it is not possible to be drier than dry! All readings above three standard deviations may be considered to be wet with obviously the higher the reading the wetter the material. Lines of equal wetness then may be plotted on the grid, see Figure 5.7.3.

False readings may occur where materials tested are composed of varying levels of naturally occurring hydrogen. The gauge measures hydrogen and assumes all the hydrogen it counts is moisture. Only if the material varies in its chemical make up does this affect the result. It is recommended that all readings are taken on a comparative basis and are not absolute.

Specification

Section 3.4 *Standards and specifications* gives general guidance.

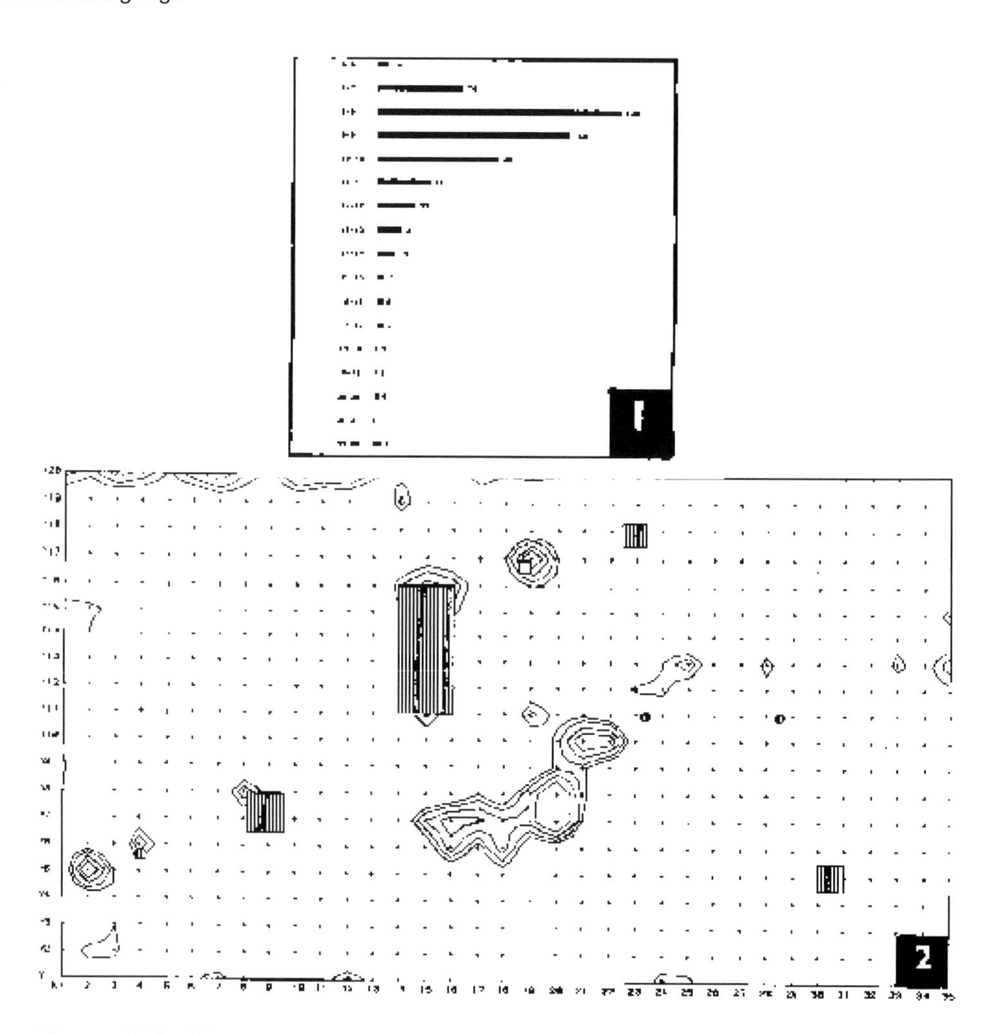

Figure 5.7.3 *Histogram and computer generated plot of data*

5.7.5 CASE STUDY

None

While no specific case study of use of the roof moisture gauge in the UK is available they are in extensive use in the USA. One extensive project where the conventional nuclear moisture/density gauge was very successfully used was for determining roof moisture on flat roofs on many properties owned by British Telecom across the UK.

5.7.6 RELEVANT DOCUMENTS

Standards and legislation

Her Majesty's Stationary Office, *Radio Substances Act*, HMSO, London, Chapter 12, 1993

Her Majesty's Stationary Office, *The Ionising Radiation Regulations*, HMSO, London, Statutory Instrument No 1333, Health and Safety, 1985 (due to be updated in January 2000)

Her Majesty's Stationary Office, *The Radioactive Material (Road Transport) (Great Britain) Regulations*, HMSO, London, Statutory Instrument No 1350, 1996

Guidance

Troxler International, *Roof moisture gauge*, Instruction Manual, trade literature

Key reading

None

5.8 Nuclear magnetic resonance

5.8.1 APPLICATION AND USE

Principle

The hydrogen nuclei in a material, principally in the moisture, behave like very small magnets that are detected using a nuclear magnetic resonance spectrometer.

Other names

None.

Application

Nuclear magnetic resonance (NMR) spectroscopy is a laboratory-based technique requiring complex and expensive equipment. It gives precise data on:

- the moisture content of a material sample
- the distribution of the moisture and
- the degree to which the water is free or adsorbed within the sample.

Usage

Materials Most materials, although the results are not reliable for plastics.

Elements Samples can be taken from any building element.

Use with

It is necessary to establish a calibration of the NMR signal for each material by oven drying and weighing. No other tests are required when samples are taken except normal recording of location etc. Other laboratory tests, such as porosimetry, can provide useful backup information.

Health and safety

As the equipment produces very strong magnetic fields, the manufacturer's instructions state that it must not be used by anyone with a heart pacemaker or similar device.

Costs and time

Preparation of a suitable sample is the most time consuming part of the process. Once the sample is prepared the moisture content can be determined within one to two minutes. Measurement of the moisture distribution takes 10–15 minutes. The marginal cost of carrying out single measurements will therefore be small, however overall charge rates may reflect the high capital cost of the equipment.

Relevance

Given a sample of appropriate size, NMR measurements allow the determination of the moisture content and distribution in a material. Tests of the change of moisture content distribution with time are used to calculate the liquid water transport coefficients of the material. The samples can be cylinders taken by coring or cut prisms with typical dimensions 50 mm × 50 mm × no limit.

As the equipment is complex and expensive and needs skilled staff to operate, it is only cost effective if measurements more detailed than simple moisture content are required.

Advantages

- NMR spectroscopy is a rapid method for determining the distribution and nature (free liquid or absorbed) of the moisture content of materials. It can be used to track changes of the distribution and measure diffusion coefficients.

Prompts and pitfalls

- A destructive test that requires skilled laboratory analysis.

Hints on accuracy

- Care must be taken when preparing samples to ensure that the moisture content and distribution are not affected.
- The most important stage of the process is in the preparation of samples so that they are representative of the material.

5.8.2 PRINCIPLES

Property

NMR spectrometers function by measuring the quantity of hydrogen atoms in a sample of interest. The hydrogen nuclei behave like small bar magnets and, when a sample is placed in a magnetic field, they try to align themselves parallel to the field, inducing a "bulk magnetisation" in the sample. The orientation of the nuclei is changed by applying radio frequency (RF) radiation to the sample, detection of the energy absorbed by the sample leads to the NMR signal that is measured. By appropriate calibration, the signal from the hydrogen nuclei corresponds to the number of water molecules present.

Test basis

The basic requirements of a system are a permanent magnet with space between the pole faces to accommodate the sample and a RF source and detector. Two techniques are available.

1. In a continuous wave (CW) system the RF radiation is swept through all the free proton resonance frequencies, giving complex information that is difficult to interpret.
2. In the more commonly used pulsed system, an intense pulse of RF radiation lasting only a few microseconds excites the nuclei. All the hydrogen nuclei in all phases are excited and when the pulse is switched off the nuclei return to their original state emitting a NMR signal as they do so.

Measurements

The NMR signal from the pulsed system is characterised by the following properties:

- the initial amplitude is proportional to the total number of hydrogen nuclei in the sample
- as the signals due to nuclei in different chemical or physical phases decay at different rates, the observed signal is a superposition of more than one component. The decay time of the signal give information about nuclei in solid, adsorbed or free liquid phases.

Units and scale

The NMR signal, which is dependant on the settings for gain, bandwidth etc, is read from a digital display or interfaced with a computer controled measurement system.

Sensitivity

The sensitivity depends on the settings of the equipment and the nature of the material present.

Accuracy

The accuracy of the calibration between the NMR signal and moisture content is dependent on the material uniformity and the accuracy of the initial material moisture content determination. The "repeatability" of the NMR signal measuring the same sub-sample of material is better that 0.1 per cent.

5.8.3 EQUIPMENT

Components NMR comprises a large permanent magnet and control box with electronics, etc.

Size The magnet is approximately 500 mm × 500 mm × 200 mm; the control equipment is a similar size.

Weight The magnet weighs 60–80 kg and the control box 10 kg.

Storage NMR is non-portable laboratory equipment.

Calibration A calibration between NMR signal and material moisture content determined by the oven-drying method must be established for each type of material if the test is being used to measure absolute moisture content.

5.8.4 METHOD OF OPERATION

Requirements

Samples Samples should be made small enough to pass through the probe head within the permanent magnet; this is usually a circular hole 40–60 mm in diameter (the cost of the equipment rises very rapidly with increasing size). There is no limitation on length.

Environmental Although no special environmental conditions are needed, stable temperatures and humidities are advisable.

Situational The high magnetic fields may effect other equipment and pose health hazards, thus a separate room for the equipment, with warning notices, is recommended.

Also as the magnet is extremely heavy (60–80 kg) it requires a robust working surface.

Human resources/skills Good technical support is needed for sample preparation and a trained technician to operate the NMR equipment.

Initial trial testing Generally an initial test is carried out with each material to optimise the settings to give the highest sensitivity and then, for moisture content measurements, to calibrate the NMR signal.

Use The NMR equipment is set up in a laboratory and a typical experimental set up for measuring liquid water diffusion rates in a prismatic sample is shown in Figure 5.8.1. Water is supplied continuously from the reservoir to one end of the sample. Scanning the NMR detector along the sample at regular intervals, builds up a series of moisture profiles that are used to calculate the diffusion coefficients.

Data recording For simple moisture determination, only the peak signal needs to be noted by the operator. For more detailed analysis of moisture distribution and state, the NMR signal as a function of time should be recorded: this is best achieved by interfacing the equipment with a computer.

Results Once the system has been calibrated, the results can be expressed directly in terms of moisture content.

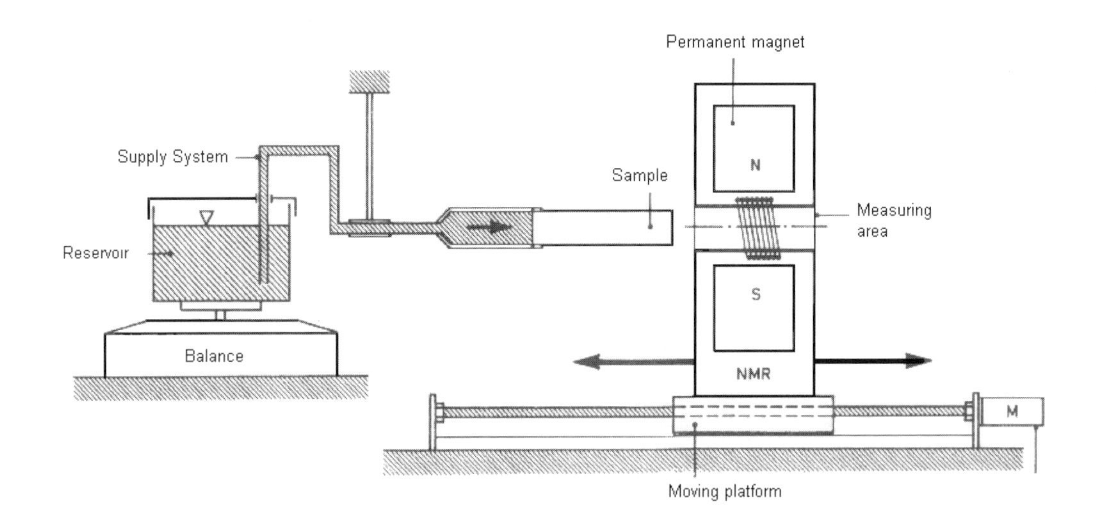

Figure 5.8.1 *Typical experimental set up (from Krus, 1995)*

Specification It is necessary to hold detailed discussions with each manufacturer to develop a specification.

5.8.5 CASE STUDY

None As the application of NMR spectrometry to building materials is relatively new, work has concentrated on developing its applicability; no case studies have been carried out so far.

5.8.6 RELEVANT DOCUMENTS

Standards and legislation No existing standards.

Guidance None available at present.

Key reading

Bruker, *An introduction to analytical applications of low resolution NMR*, Bruker minispec application note 3, 1994

Pel, L, *Moisture transport in porous building materials*, Eindhoven University PhD thesis, 1995

McDonald, P and Strange, J, *Magnetic resonance and porous materials*, Physics World Vol 11, pp 29–34, July 1998

Krus, M, *Moisture movement and transport coefficients of porous mineral building materials – new measuring techniques*, University of Stuttgart PhD thesis, 1995

5.9 Oven drying method

5.9.1 APPLICATION AND USE

Principle

The method is based on accurate measurement of the loss of weight of a representative sample of the material following controlled oven drying.

Other names

Gravimetric method.

Application

The oven drying method provides a laboratory determination of the dry or wet weight moisture content (MC) of a sample of a material from a specified initial moisture condition; for example as received. This method of moisture content determination has been used for many years and provides an accepted base measurement in the determination of the moisture condition of a material. It also provides a measurement against which moisture readings from generally non-destructive testing (NDT) techniques can be compared and validated: when making such comparisons it is important to understand the likely differences between the two measurements (see Section 3.3 *Moisture measurements*).

Alternative site-based methods of drying samples quickly have been used, mainly for aggregate and soils testing, based on warm-air drying units, industrial microwave ovens and infrared heaters. These drying methods are generally used where large numbers of samples are being tested and the result is critical to the site process. The method requires to be validated against controlled temperature oven drying for each type of material. Extra care must be exercised when using these methods as:

- the temperature is uncontrolled
- over heating the sample may cause particle disintegration and lead to direct loss of material and potential injury to the operative
- free moisture must always be available to prevent permanent damage to the microwave oven
- building sites do not generally provide the controlled environment conductive to the accurate weighing of small samples.

In either respects, site-based methods are similar to laboratory methods.

Usage

Materials

Brickwork and blockwork
Mortar and render
Plaster
Concrete
Stone
Timber,
Aggregate and soil.

Elements

All elements.

Use with

This method can be used in conjunction with the other test methods to determine the moisture condition of materials. It may be particularly useful in confirming a site reading for moisture content taken with a calcium carbide moisture meter, or following detection of damp areas with an electrical capacitance or resistance meter.

Health and safety

Personal protective equipment (PPE) needs to be worn particularly if the material type is an unknown, as well as protective gloves for the hands and forearms when placing or removing samples from the oven. After removing samples, care must be taken to ensure that the hot sample and container are placed in a safe and well-marked area to cool.

Exercise due care when handing and using chemicals, eg for hygroscopic moisture content (HMC) tests. Check COSHH data sheets.

Costs and time

Make due allowance for the time, and the associated costs, needed for:

- taking samples of materials on site and transporting them to the laboratory
- processing in the laboratory particularly drying/cooling times to achieve a constant weight, a minimum of two hours for small volume powder samples and 24 hours (or days) for larger lump samples
- any pre-conditioning of the sample that may take days for example for HMC tests.

Relevance

Oven drying determinations of moisture content are often required as part of the quality control procedure to confirm the moisture condition of materials prior to use as determined by NDT techniques.

MC and HMC may be needed to conclude the assessment of moisture levels within the body of an element which authoritative sources and research has shown as vital for the correct diagnosis of dampness problems. The HMC tests are used to help demonstrate the presence (or absence) of hygroscopic salts (chemical tests provide additional data).

Advantages

- An accurate laboratory based method that is simple, reliable and repeatable.
- As the sampling method to obtain a representative sample generally requires more material to be collected than required for moisture content determination this makes provision for other laboratory testing to be carried out eg (chemical) or HMC tests.

Prompts and pitfalls

- A destructive test: when sampling material from the building fabric the requirement will generally be to minimise the number of samples and volume of material extracted. In some cases sampling will be unacceptable and alternative test method(s) used.
- Materials other than water may be driven off.
- Samples need to be carefully and correctly handled during collection, transportation and processing.
- Normal oven drying of samples at 105°C may not be appropriate for all materials. For example, at 105°C retarders used within gypsum plaster may be affected by heating and the loss of such material may influence sample weight and hence the measured moisture content. It is therefore recommended that a temperature of less than 40°C be used for calcium sulfate and 45–55°C for calcium sulfate (gypsum) plaster. Alternatively, a differential drying technique to achieve constant weight at initially 45°C then at 200°C will enable the determination of physically and chemically bound water respectively. (In the manufacture of gypsum plaster variations in heating from 150°C to over 190°C may occur and it will require such temperatures to alter the plaster chemically).

Hints on accuracy

- The volume of the container should be carefully selected to minimise moisture transfer between the sample and remaining air space and to minimise the quantity of material likely to adhere to the container surfaces. Keep the samples at a reasonably constant temperature from the time of sampling to testing.
- Sample containers must be airtight.
- Reliable weighing equipment is required to ensure the quality of the calculated result.
- Ensure that the sample has been dried to a constant weight and that the sample has cooled sufficiently to permit an accurate weight determination (and not damage the balance).
- When assessing the MC of hygroscopic materials such as in timber the sample should be placed in a dry environment to cool following oven drying at 105°C to prevent absorption of moisture from the air before weighing: eg in a desiccator containing a desiccant such as silica gel.
- When undertaking HMC determination the time period that the sample spends in artificial relative humidity conditions prior to oven drying should be at least two days rather than the hours suggested in some publications.

5.9.2 PRINCIPLES

Property Careful oven drying can drive off the moisture from a sample of material.

Test basis The moisture content of a sample material is determined by measuring the loss of weight through oven drying.

Measurements A simple formula is used for obtaining the sample moisture content from the recorded weights:

$$\frac{100 \text{ x } (W_w - W_d)}{(W_d - W_o)} = \text{percentage dry weight moisture content}$$

$$\frac{100 \text{ x } (W_w - W_d)}{(W_w - W_o)} = \text{percentage wet weight moisture content}$$

where W_w = weight of damp sample and petri dish
W_d = weight of dried sample and petri dish
W_o = weight of the empty petri dish.

Units and scale The unit of measurement is gram (g) and the balance should be selected to have a measurement range appropriate to the sample size and/or required accuracy.

Sensitivity The sensitivity of the balance will be dependent on the measurement range, which is typically better than 1 in 10 000. Modern small capacity electrical balances can discriminate to better than 1 mg (0.001 g).

Accuracy A well-maintained balance will normally be accurate to ±1 division over the full measurement range. When converting weight to moisture content, the combination of the weighing accuracy and the sample weight will generally determine the accuracy in percentage terms.

5.9.3 EQUIPMENT

Components

A ventilated oven capable of sustaining 105°C ± 2°C conditions **(the temperature may vary for different materials).**

An accurate balance with a scale graduations/full-scale reading appropriate to the sample size and/or accuracy required.

Petri dishes or similar.

A desiccator or other relative humidity chamber.

For sample extraction, a low-speed electric drill and drill bits or, if appropriate, a club hammer and hard chisel for lump samples.

Airtight containers for removal of site samples to the laboratory.

Size

Ovens vary from small 50 litre capacity units to free-standing 1000 litre capacity cabinets.

Balances can have weighing capacities from 200 g to 50 kg or greater.

Weight

Ovens weigh from 40–200 kg and balances 3–20 kg.

Storage

The weigh balance should be kept in a clean, dry environment and ideally not be subject to wide temperature variations.

Calibration

The oven should be well maintained and periodically checked to ensure that the temperature and temperature control are within the required specification.

Maintenance and regular calibration is essential for weigh balances. Calibration should be carried out using a wide range of size of weights, which are traceable to National Standards, and cover the full-scale measurement range.

5.9.4 METHOD OF OPERATION

Requirements

Samples

Testing is usually on the basis of single samples although some processors prefer to test samples in duplicate as an accuracy check. For guidance on sample weight see Section 3.3 *Moisture measurements*.

If testing for both hygroscopic moisture content and moisture content sufficient sample should be collected to undertake both tests and divided, using a recognised technique to ensure each sub-sample is representative.

Environmental

The laboratory should have a controlled environment.

Sample extraction is generally from inside the property. When sampling outdoors, ensure that rain does not contaminate the sample.

Situational The area used for weighing should ideally be draught-free and have adequate space for the operator without obstructing access routes. Adequate space should be provided for the storage and processing of the numbers of samples expected through the laboratory.

When sampling, enough space is required to operate the extraction equipment and collect the material.

Human resources/ skills The technician needs to recognise the importance of carrying out the procedures methodically to ensure repeatability and that samples need to be correctly handled during collection, transportation and while processing.

Initial trial testing A test programme may need to be undertaken to demonstrate the effect of oven temperature on the weight loss for some materials, ie it may be necessary to dry the sample for a longer period at a lower temperature to ensure that only free moisture is driven off.

Use

The procedure comprises site sample extraction and laboratory processing.

Site work:

- when extracting the sample of material from the building fabric use a low speed electric drill. Drill for short periods so that heat does not build up and dry out moisture; a 9 mm drill bit is recommended for mortar (BRE Digest 245) to minimise evaporation losses through friction. Discard the material from the first 5 to 10 mm to minimise contamination from surface moisture. Record the depth over which the sample is extracted
- endeavour to extract a sample of a single material ie prevent mixing of materials
- transfer the sample immediately into an airtight container and seal to minimise the risk of moisture changes
- clearly label the container ready to transport the samples to the laboratory.

Laboratory work for MC on powdered samples:

- check that the balance is set to zero and weigh the petri dish
- thoroughly mix the sample by shaking it in its airtight container. Spread the required weight of thinly on to the dish for oven drying
- weigh sample and dish immediately before carefully placing in oven to dry
- after cooling re-weigh the sample and dish. (If the sample is hygroscopic it will need to cool in a desiccator over a desiccant to prevent moisture absorption from the air)
- return sample and dish to the oven for a further period of drying; the dry weight may need to checked a number of times to ensure that it remains constant before the final reading is recorded. When this operation has been repeated a number of times for a given material condition and sample size the drying time should be known and the procedure can be simplified.

Initial **laboratory work** may require the sample to be pre-conditioned in a sealed container at a known temperature and relative humidity to achieve a constant weight; the first part of a hygroscopic moisture content (HMC) test (see Section 3.3 *Moisture measurements*). The air over a saturated solution of common salt (NaCl) in an enclosed container will provide around 75 per cent relative humidity at 20°C. A range of chemicals can be used to produce saturated solutions for varying relative humidity levels (eg potassium chloride 85 per cent, potassium nitrate 93 per cent both at 20°C).

Data recording

For all MC determinations the weights will require to be tabulated and the percentage values calculated using the appropriate formula. The condition of the sample at both the start and end of the test are important and should be recorded. For example at the start the sample will generally be "as received" or it may be pre-conditioned in controlled conditions (HMC test). The end point will generally be oven dry ie dried to a constant weight at a given temperature or, for typically sand and gravel, may be a surface dry condition determined by visual inspection.

It is essential that sufficient data be recorded to enable the test to effectively be repeated ie sample location and conditions at time of sampling through to the full test procedure as carried out.

Results

The "as received" MC results when oven-dried will generally relate to the free water content (water absorbed by capillary action and by atmospheric absorption) and are equivalent to the site readings taken with the calcium carbide moisture meter.

Specification

Do not be over-prescriptive, the need for oven dry MC measurements for particular situations are probably best left to the discretion of the testing house, see Section 3.4 *Specification and standards* for further general guidance.

5.9.5 CASE STUDY

Rising damp

In a court case, of alleged negligence of a remedial dpc contractor to detect rising dampness, the oven-drying method and calcium carbide moisture meter were used to help determine the source of moisture.

Construction A solid external walled detached house.

Reason The prospective purchasers of the house had received a building society surveyor's report that suggested that a limited amount of rising dampness had been traced whereupon the remedial dpc contractor was invited to inspect.

Following the inspection the contractor suggested by telephone to the selling agent that although some isolated readings were found on site they were not sufficient evidence to conclude for the need for a dpc, clearly an honest opinion as this was against the contractor's own commercial interest.

On taking occupancy the new owners experienced a visual picture on the wallpaper of the kitchen that was the epitome of rising dampness and instructed their solicitor to proceed against the contractor.

Results The reports in support of the owner claim of negligence were based only on visual and electric moisture meter readings. The assessor for the contractor adopted a four-stage methodology that included:

- a visual inspection
- electric moisture meter assessment of the internal surfaces of the wall
- extraction of wall materials samples for site carbide testing
- oven drying testing and a salt analysis of the wall surface (wallpaper and plaster).

Success The MC results from the carbide and oven drying concluded low body moisture contents in the wall (around 1 per cent). The HMC results were all higher than the MC, and the salt analysis concluded a strong presence of hygroscopic salts.

Assessment The wall, at some stage, had been affected by rising dampness with a long-term accumulation of associated salts. However, the testing was successful in concluding that despite the salt presence the rising dampness was no longer present. (Site observations of the outside of the property concluded that the ground levels surrounding the building had been substantially lowered and this appears to have had a desirable effect on the rising dampness.)

The remedial dpc contractor could not be held negligent for failing to detect rising dampness if the dampness was no longer present. Hygroscopic salts were causing the face dampness to the wallpaper and the solution was to replace the wallpaper and plaster coverings to the wall.

5.9.6 RELEVANT DOCUMENTS

Standards and legislation

British Standards Institution, *Specification of aggregates from natural sources for concrete*, BSI, London, BS 882: 1983

British Standards Institution, *Specification for timber*, BSI, London, BS 1186: Part 1: 1991

British Standards Institution, *Performance requirements for electrically-heated laboratory drying ovens*, BSI, London, BS 2648: 1955

British Standards Institution, *Code of practice for installation of chemical damp-proof courses*, BSI, London, BS 6576: 1985

British Standards Institution, *The performance of damp-proof courses installed to prevent rising damp*, BSI, London, DD 205: 1991

Guidance

Building Research Establishment, *Rising damp in walls: diagnosis and treatment*, BRE, Watford, Digest 245, minor revisions ed.1986, reprinted 1989

Key reading

Johnasson, C H and Persson, G, *Moisture absorption curves for building materials* National Research Council of Canada, Technical translation 747, Ottawa, 1998

Further reading

Howard, C A, *An evaluation of the techniques employed to diagnose rising ground moisture in walls*, Liverpool Polytechnic, MPhil thesis, 1986

5.10 Radar

5.10.1 APPLICATION AND USE

Principle

Radar is a tool for measuring variations (not necessarily absolute values) in the moisture content of structures over large areas. The speed, strength and frequency content of radar signals can be measured in order to identify changes in moisture content.

Other names

Impulse radar, ground penetrating radar, sub-surface radar.

Application

Radar is used to determine construction detail, variation in condition (quality control) and moisture content.

Usage

Materials

Concrete
Brickwork and blockwork
Masonry
Screed
Timber
Aggregate and soils
Asphalt and other non-conductive roofing materials.

Elements

Walls
Ground floors
Roofs
Structural frame
Stairs.

Use with

In many applications it is essential that radar data be calibrated using other methods in order to determine what variations in the data are related to changes in moisture content. Ideally samples should be removed for oven drying but failing that, methods such as electrical resistance and capacitance meters can be useful as they also measure the variables that effect the transmission of electromagnetic waves (radar).

Health and safety

There are no particular heath and safety issues associated with the transport or use of this radar equipment. Radio emissions are comparatively low powered and present none of the hazards often associated with radiography (X-ray). The equipment is therefore suitable for use in occupied buildings.

Costs and time

The costing of radar investigations are typically calculated in a number of ways including fixed lump sum, per metre or per day, depending on the type and size of structure under investigation. On a structure such as a floor slab (free of obstructions), an area of 1000 m^2 could be surveyed in one day.

In the majority of applications it is access considerations that limit the area covered in a days surveying. This is demonstrated by the fact that on structures such as large factory floors and roads where the equipment can be operated from a vehicle it is possible to survey approximately 3000 m length of single carriageway in some detail in one day.

A full day of data collection may require around five days of processing and analysis.

Relevance

Radar is most suited to mapping variations in the bulk moisture content across extensive uniform structures such as walls and floors and when other construction or condition information is required.

Advantages

- Radar provides a totally non-destructive test method.
- The equipment is light and generally very portable.
- It allows comparative assessment of variations in moisture content over large areas for relatively low cost per unit area.
- Radar supplies a range of other construction and condition information that can often be used to identify the mechanism of moisture ingress, such as structure thickness, relative compaction of materials, reinforcement layout, reinforcement corrosion and delamination.
- It can give some idea of the depth of the moisture if it is contained within resolvable structural elements
- Radar has a greater penetration than the majority of other methods, up to approximately 1 m depending on the situation.
- Digital recording of data allows easy comparison of changes over time with a number of site visits.

Prompts and pitfalls

- Radar is ineffective on metallic and highly reinforced concrete structures (bars at centres closer than approximately 100 mm).
- As with the majority of in-situ tests it is generally not effective as a means of obtaining exact measurements of moisture content.
- Radar requires in-depth knowledge of the technology to be effective.
- Radar cannot be used on new (less than two months old) concrete and mortar due to its high conductivity.
- The equipment is expensive.
- It often requires calibration by other techniques.
- It does not give instant results; data requires post processing and analysing by specialists.

Hints on accuracy

When commissioning a radar survey the client will need to ensure that:

- all relevant information about the site, the construction materials and history are available prior to the site work
- as far as possible, the survey methodology is planned prior to the site work
- all data is accurately attributed to the structure under investigation
- any calibration data, such as measured depths or point tests from other techniques is accurately related to the radar data.

5.10.2 PRINCIPLES

Property The effects of variations in the electrical conductivity and dielectric constant on the transmission of electromagnetic waves are measured in order to identify changes in moisture content.

Test basis Radar involves the transmission of short duration pulses of radio energy into a structure and the measurement of the reflected signals. In the majority of applications the transmitter and receiver are placed side-by-side in a unit referred to as an antenna (see Figure 5.10.1).

In most applications the electrical conductivity and dielectric constant of the material largely determine the behaviour of a radio wave during its transmission through a material. The conductivity determines the rate at which energy is lost in the transmission through the material, and the dielectric constant determines the velocity of transmission and the percentage of energy reflected and transmitted on at a material boundary. The energy loss caused by increased conductivity is frequency-dependent, ie there is more energy loss in the higher frequencies.

Moisture contained within building materials is generally conductive as a result of the presence of dissolved ions. It can therefore be identified by increased loss of radio energy especially in the higher frequencies. Additionally moisture has a much higher dielectric constant than most building materials and can therefore also be identified by decreases in the velocity of transmission.

Measurements The velocity and frequency characteristics of the reflected radio pulses are recorded (see Section 5.10.5, Figures 5.10.3 and 5.10.4).

Units and scale The lapsed time scale can be related to distance travelled by the antenna ie to the position on structure. The frequency, amplitude and pulse time responses are usually "interpreted" from the measurement scale, ie treated comparatively, but can be "calibrated" to the desired units if required, eg moisture content.

Sensitivity In the majority of situations radar is sensitive to variations in moisture content in the order of 1 per cent.

Accuracy The accuracy of radar for the determination of moisture content is largely determined by, and is therefore a function of, the accuracy of the calibration method.

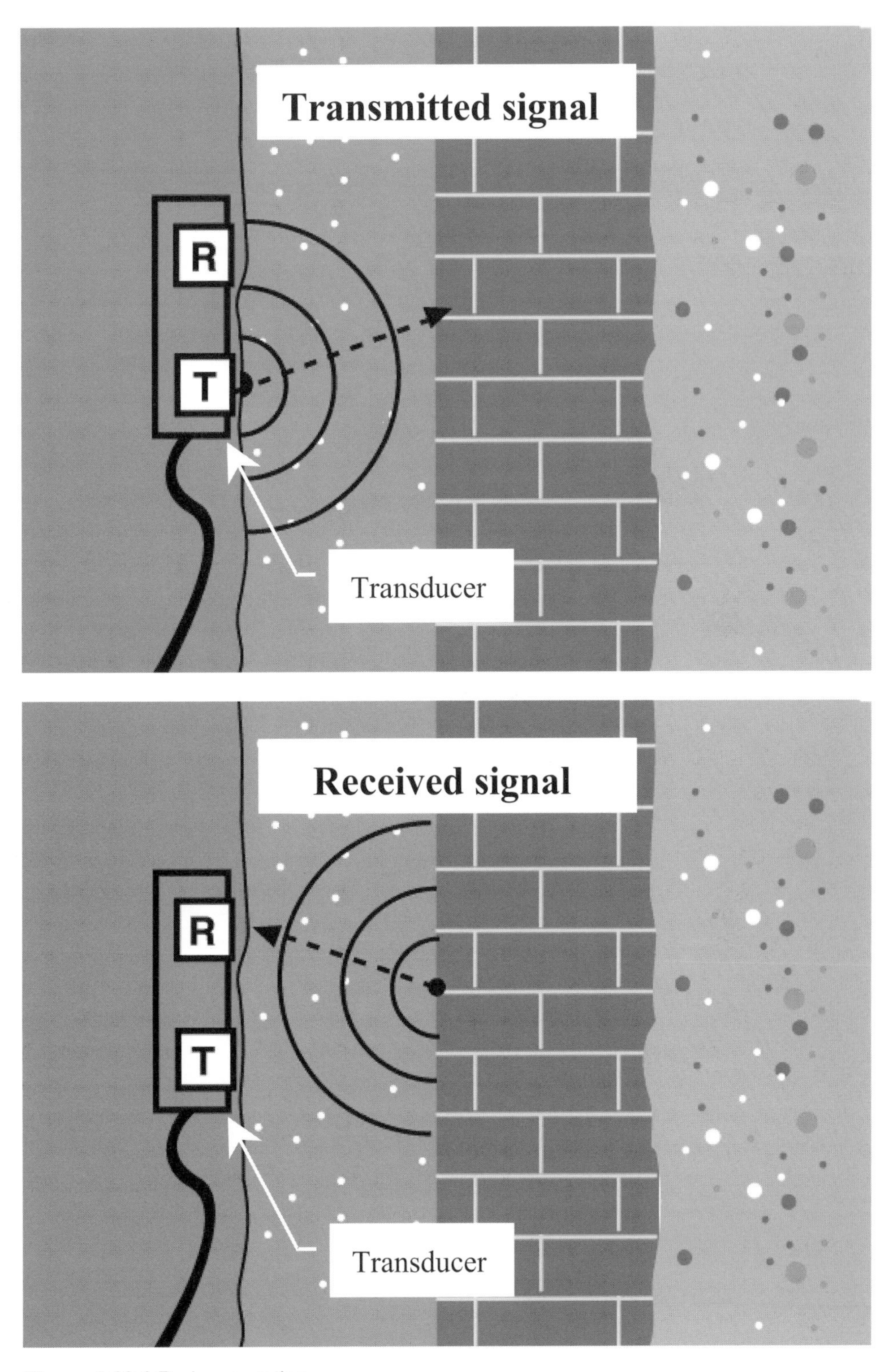

Figure 5.10.1 *Radar operation*

5.10.3 EQUIPMENT

Components A typical radar set-up, as shown in Figure 5.10.2, comprises a control unit powered by a 12 V battery, a printer and a transducer. The transducers come in a range of centre frequencies extending from approximately 15 MHz to 2.5 GHz. Higher frequencies give better definition but reduced penetration and lower frequencies give better penetration and reduced definition. The higher end of the frequency spectrum (400 MHz to 2.5 GHz) is generally employed in the investigation of buildings and other civil structures and the lower end is used in ground investigations.

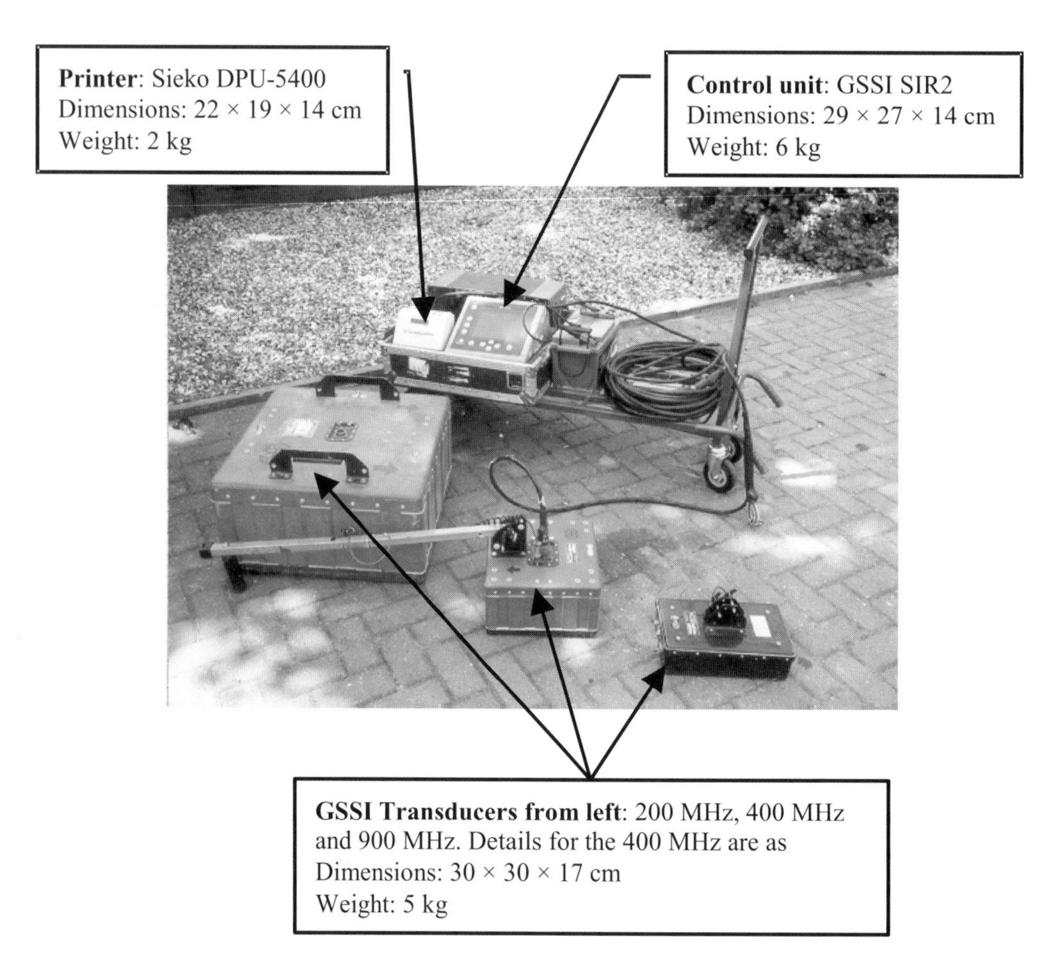

Figure 5.10.2 *Typical radar equipment*

Size The majority of commercially available equipment is small enough to be carried up and down ladders etc but must be treated with some care since it may contain delicate electronics.

Weight 5 kg.

Storage The equipment should be stored in a dry environment.

Calibration Radar data generally needs to be calibrated with other methods such as sampling for oven drying, carbide moisture meter or electric meters (capacitance or resistance).

5.10.4 METHOD OF OPERATION

Requirements

Samples The antenna is passed over the surface of the structure under investigation at a controlled speed and scans are taken at regular intervals. The scan rate in most radar systems today is such that the collected data effectively forms a continuous cross-section, enabling assessment of a number of features including the construction, condition and moisture content.

Environmental The construction of the majority of radar systems makes them susceptible to damage from heavy rain.

Situational Radar functions most effectively when the transducer is in close contact with the structure under investigation; a smooth surface is therefore preferable (for increased accuracy) but not essential

Radar data can be more easily analysed when it is collected in long continuous profiles; clear uninterrupted access to large areas of the structure under investigation is therefore desirable.

Radar is more effective on structures of uniform thickness.

Human resources/ skills Specialists with expertise in radar and a good knowledge of construction practice usually conduct radar surveys.

Initial site testing It may be advisable to undertake initial testing to be sure that the radar will produce satisfactory results

Use

The stages of a typical site survey are as follows:

- determine the extent and relevant dimensions of the structure under investigation
- use this data along with any provided prior to the site work, such as the expected extent of the damp area, to confirm the radar set up parameters
- depth of search; this also affects the centre frequency of the selected transducer(s)
- spacing and position of the radar profiles
- use the dimensions determined above along with marking materials such as chalk and tapes and/or survey wheels to mark out an orthogonal grid over the area under investigation. This grid must be referenced back to the structure under investigation
- set up the radar equipment with the selected transducer(s) and depth of search
- use the orthogonal grid to gather data over the selected profile locations
- use the radar data along with any other relevant information to determine the location of any calibration points and carry out the calibratory tests.

Data recording

Some of the digital radar systems available today allow data from the collected profiles to be recorded simultaneously onto a printer for immediate on-site analysis and a hard disk for post processing.

It is essential that the location of all of the collected radar profiles be recorded along with any other data to be used, for example, for calibration purposes.

Results

The collected data can be analysed in a number of ways to map areas of increased moisture content, depending on the type of structure and quality of the collected data. Increasing the number of methods of analysing the data for any particular structure increases the confidence of the result. Until the fairly recent introduction of digital radar systems, certain methods such as frequency analysis were not available so the confidence levels associated with projects were generally lower.

The range of suitable methods of analysis available in any particular situation is governed to a large extent by the uniformity of the structure under investigation.

For structures of uniform thickness such as brick walls, where the radio signal from the back of the structure can be resolved, the following options are available:

- map changes in the two way **travel time** of the signal from the back of the structure
- map changes in the **frequency content** of the signal received from the back of the structure
- map variations in the energy of the signal received from the back of the structure by using the **amplitude** of the signal.

For structures of variable thickness, where the back of the structure cannot necessarily be identified, the following options are available:

- map changes in the **frequency content** of either the whole or part of each of the traces
- if a material boundary can be resolved at a variable depth it may be possible to map variations in the **speed** of the radio signal by measuring the two-way travel to a fixed depth with a number of different transmitter/receiver separations (this technique is referred to as common mid-point, CMP)
- if a material boundary can be resolved at a variable depth it may also be possible to map moisture variations by examining the **amplitude** of the received signal (allowance might have to be made for the variable depth)
- for materials with a number of point reflectors such as gravel a general loss in received signal **strength** can be used to map areas of increased moisture content (This is a particularly vague technique and must be used with a great deal of care).

It should be noted that radar is affected by the free moisture and does not distinguish between capillary moisture and hygroscopic moisture. The methods that are suggested in BRE Digest 245 for distinguishing between causes of damp could be used to make the distinction.

The results of investigations can be presented in a number of different ways including plans, elevations and/or sections with contours showing variations in the moisture content. In situations where it has not been possible to calibrate the radar data the results may be in the form of unit-less contours of increasing moisture content.

On some uniform thickness structures such as brick walls variations in the dielectric constant of a material can be translated, with some additional data such as the void ratio, into percentage moisture content.

Reporting requirements are discussed in general in BRE Report No 340.

Specification

The specification of a radar investigation should state as clearly as possible what is required in terms of the following:

- information to be supplied prior to the site work
- the site work
- the final report.

Due to the complexity of radar the majority of persons specifying work will need to rely to a certain extent on the radar contractor to specify the optimum types of transducers and radar profile spacing for the site work. A list of approved radar operators can be obtained from the trade association of radar users, EuroGPR.

It should be noted that the outcome of radar investigations could sometimes be inconclusive so it may be necessary to specify a trial before the main contract.

Extensive guidance on the specification of radar surveys is also given in BRE Report No 340.

5.10.5 CASE STUDY

Cavity wall

Construction The walls comprised a brick external skin and block internal skin tied with butterfly wall ties.

Reason The cavity walls of recently constructed houses were investigated as a quality control measure.

Radar with a 1.5 GHz transducer along with a pulse induction metal detector was taken on site as the main investigative methods. Once on site it was immediately apparent that the metal detector would not locate the wall ties since they were stainless steel. Radar was therefore the main investigative method used.

Results Figure 5.10.3 indicates the location of wall ties set in dry brickwork. The location of all of the wall ties identified was, as can be seen on the photograph chalk marked directly onto the wall in this case.

Figure 5.10.4 indicates the effect of increased moisture levels on both the two way **travel time** and **frequency content** of the radar data. Amplitude variations were inconclusive in this case.

Success Radar successfully identified the pattern of wall ties and variations in the moisture content of the wall.

Assessment The variations in moisture identified in the radar data were clearly corroborated by visual inspection.

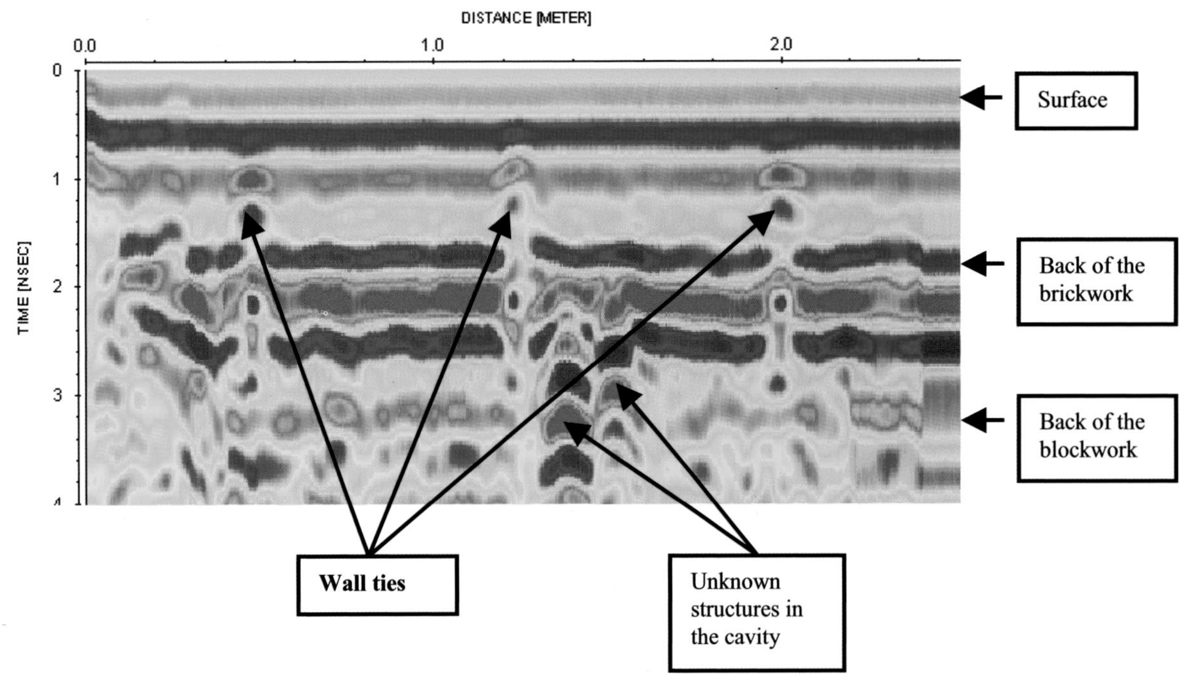

Figure 5.10.3 *Conclusive radar response*

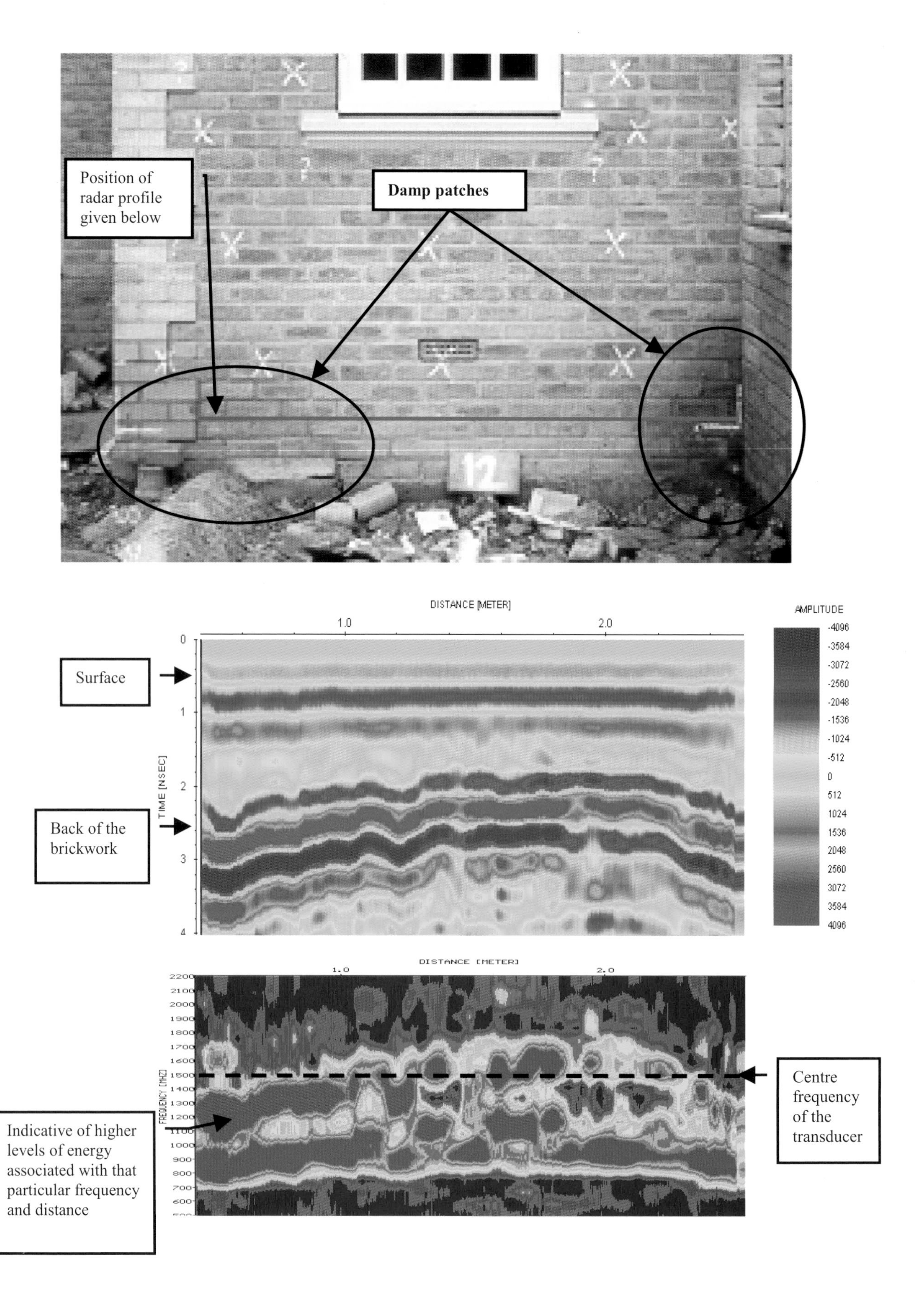

Figure 5.10.4 *Inconclusive radar response*

5.10.6 RELEVANT DOCUMENTS

Standards and legislation

None.

Guidance

Matthews, S L, *Application of subsurface radar as an investigative technique*, CRC Ltd, London, BRE Report BR340, 1998

Concrete Society, *Guidance on radar testing of concrete structures*, Concrete Society, Slough, Technical Report No 48, 1997

Her Majesty's Stationary Office, *Use and limitations of ground penetrating radar for pavement assessment,* Highways Agency, HMSO, London, Design manual for roads and bridges, Volume 7, Section 3, Part 4, HA 72/94, 1994

Annan, A P, *Ground penetrating radar workshop notes*, Sensors & Software, private communication (available from the manufacturer)

Key reading

Forde, M C, McCavitt, N, Binda, L and Colla, C, *Identification of moisture capillarity in masonry using digital impulse radar*, Structural Faults & Repairs, 1996

Useful contact

EuroGPR, tel: 01788 536389

5.11 Thermographic inspection

5.11.1 APPLICATION AND USE

Principle

Thermographic inspection measures the heat being radiated by a surface and from the variations in the radiation detected the presence of moisture may be inferred. It does not measure moisture content.

Other names

Thermography, thermal imaging, infrared thermography. **Note** that false colour infrared, infrared photography and night vision are **not** the same method.

Application

The use of thermography to identify heating or insulation problems is well known. This application concerns its use to examine the variations in the thermal emissions of masonry, concrete and wooden structures arising from changes in their condition and thermal performance as a result of the presence of moisture within the fabric. Examples of appropriate applications are:

- locating leaking water bars and cracks in slabs
- identifying the spread of damp through a wall from a localised source
- identifying a leaking joint in a projecting cornice can be located with ease from ground level, minimising the need for expensive access equipment.

The process is essentially comparative, and should not be considered as able to give absolute values of moisture.

It is the only method that can observe relative moisture content from a distance, requiring only line of sight to the surface to be studied. It provides sufficiently repeatable measurements to allow its use in long-term monitoring.

It is a common misconception that thermal cameras measure the surface temperature of objects. However this is **not** the case. The camera measures the radiation from the surface. The surface, its material, conditions and surroundings are therefore crucial to the effectiveness and information recovery of the technique.

Usage

Materials

Porous materials, such as:

- concrete
- masonry
- brickwork
- timber,

preferably without weatherproofing paint finishes, although paints are suitable provided the finish is complete and in good condition.

Impervious surface materials such as metal and many thermoplastics are inappropriate.

Elements

Any plane or reasonably constant planar surface, eg walls, floors, roofs, but may include enrichment and detailing such as cornices, pilasters or decorative plasterwork.

Use with

In many applications it is essential that the testing include observation of the following, before assessing or analysing the results of the application:

- the thermal environment – sunlight, heating systems, potential thermal reflectors and emitters
- the physical environment – wind speed and eddy currents, natural and applied surfacing eg paints, algae, lichens dust/particle deposits
- any other physical effects .

An estimate is first made of the likely effect of moisture in the object, which is then tested against reality and calibrated using any other appropriate method, including simple observation. Changing environmental conditions should have a fully predictable effect, eg if the observation of thermal difference depends on evaporative cooling, the contrast should generally decrease with lower wind speeds.

Health and safety

No special hazards or risks are involved, other than those which normally apply to work on construction sites and other work places. Most commercially available systems are now entirely electrically driven, using either an internal compressor to chill a "cold" sensor or a heater unit to stabilise a "warm" sensor. Electrical power requirements for the systems are mainly battery operation and are not onerous.

Some older models may require a supply of liquid nitrogen (to chill the sensor) that will be transported around the site. There are many sensitive environments where such a supply will be unsuitable, however, and more modern equipment preferred.

Particular care must be taken to minimise the risks to personnel when carrying out nighttime investigation, eg adequate lighting to access routes.

Costs and time

Frequently, when working on external surfaces, only a small time window may be available during the course of a day. During this time a very large area may be covered once the correct conditions have been established. It may equally be necessary to wait for the optimum weather conditions, which may delay the start of the investigation or force a break in the survey progress. In the UK there is typically a seven- to ten-day cycle between cyclonic and anticyclonic conditions, thus any schedule for an exterior survey should contain such an allowance.

Careful consideration should be given to the final use of the images collected. If the task is tracking a leak in a floor slab during a repair, the images may only be needed as a background reference. If they are to be used in the study of facing stones of an ancient monument, however, there may be a need to consult them frequently, to not only determine damp but also precisely which stones will need repair.

The reporting requirements, time scale and costs therefore vary significantly with the purpose of the survey. For example, 1000 m^2 of basement floor slab and edge kickers could be surveyed in one day for leaking joints and cracks, and would be independent of weather conditions. However, on an external façade, a continuous enriched ashlar facing 75 m long and at least 7 m high could be covered in the same time.

Relevance

Thermographic inspection is most appropriate when:

- the relative moisture content across a large area needs to be found quickly and economically
- access to the particular surface cannot be readily achieved and it is advantageous to locate and consider the moisture content only in limited areas
- there is an opportunity for quality control and whole-life monitoring of large developments.

Advantages

Thermographic inspection provides:

- remote access
- instant response, therefore rapid scanning of large areas
- visual referencing
- relative moisture content assessment
- light and generally very portable equipment.

Prompts and pitfalls

- Careful pre-planning to ensure that conditions are optimised will almost always increase the value of an investigation using thermographic inspection. It is a popular accusation that an instrument "did not work", which leads to the false conclusion that the technique "does not work", when all that has happened is that it has been inappropriately applied.
- The equipment is deceptively easy to use, being comparable in form and arrangement to domestic visible light video systems, and producing images which can be readily related to the visible world but with control of temperature level and range, rather than light intensity. Most failures to make the best use of the technique are the result of recording the obvious and larger-scale thermal variations, which may be completely outside the purpose and interest of the task.
- It should be recognised that thermographic inspection:
  - responds to the surface emissions only
  - is ineffective on reflective surfaces such as metals or standing water
  - is inappropriate for absolute measurements of moisture content
  - imaging systems are expensive and are best used, and the images analysed, by experienced personnel
  - is sensitive to the ambient thermal environment and wind
  - surveys often need to be planned around weather, time of day and cyclic thermal events
  - may require night-time working for optimum imaging conditions
  - may produce thermal differences that have more than one physical cause, which includes moisture.

Hints on accuracy

- Thermographic inspection for moisture generally assumes a mechanism by which the moisture will be identified – evaporative cooling, heat transportation etc. On site another mechanism may override this one. It is important that the forward planning of the site work prepares for this, and that the survey plan is altered accordingly.
- A pilot study prior to the main survey is frequently the best way to optimise the outcome, provided it is justified by the survey extent.
- As with most expensive technology, using an experienced practitioner will increase the chances of acquiring accurate and meaningful results.

5.11.2 PRINCIPLES

Property

All bodies, even cold ones, emit "heat" in the form of infrared radiation. This radiation can be detected by thermographic cameras, which in effect measure the temperature of a target surface with considerable accuracy. A damp surface tends to be cooler than surrounding dry areas as a result of heat being lost through the process of evaporation. Thermographic cameras detect this difference primarily by comparison with the dryer areas visible in the same image, literally yielding a picture in which suspect areas are conspicuous.

Thermal radiation

Any object above absolute zero (-273°C) emits thermal energy and it is this energy that can be captured and measured. It is measured across a spectrum from:

the "near" infrared or short wavelength bordering on the red end of visible light
to the "far" infrared or very long wavelengths.

The "hotter" the object, the higher the energy and the shorter the wavelength, so a building at -10°C emits enough radiation to make measurements.

The thermal envelope of a building continually absorbs energy from its surroundings. The sun, the atmosphere and human activity all contribute and a separate "heat source" is therefore not necessary; the actual emissions of an object depend more on the materials of the surface and the surrounding conditions.

Provided that the material forming each element of the envelope does not vary, eg it is made entirely of one type of brickwork, the thermal radiation will be constant. Where a single material shows variation in its surface temperature without an apparent reason, such as surface distress or a localised change in the heat flow into or out of the material, then another physical cause must be at work.

Moisture will alter the specific heat of an object so it does not heat up and cool down at the same rate as a dry material. Hence, as the environment changes, evaporative cooling will create a temperature differential where a moisture gradient exists, especially if there is air movement.

Test basis

The basic components for a survey, as shown in Figure 5.11.1, are the thermal imager, a viewer to allow on site assessment of data and a recording system (advantageously both a continuous analogue video and a digital still store). All these require an adequate power supply, either batteries or mains.

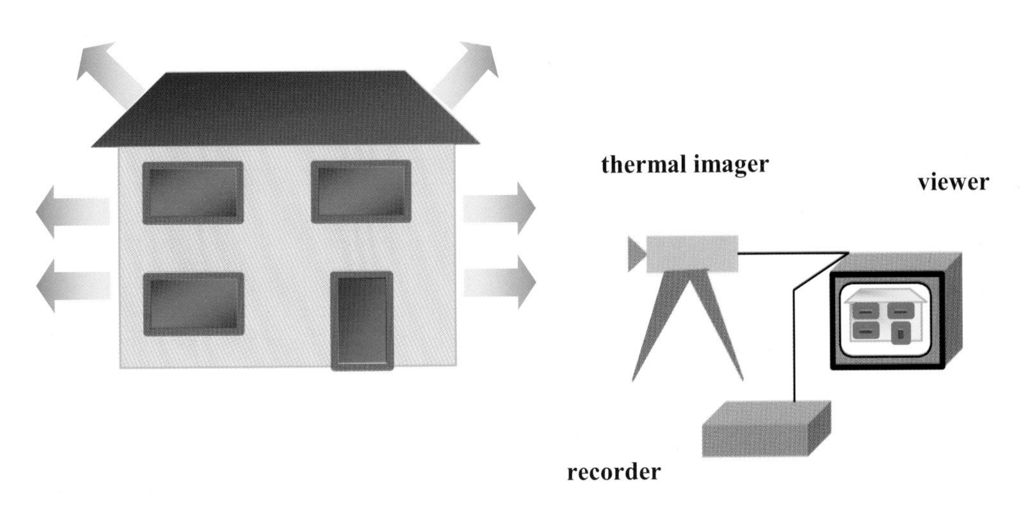

Figure 5.11.1 *Schematic view of basic components for a thermographic inspection*

Measurements Most systems will have an operator-controlled conversion for emissivity. This can be corrected by measuring the actual surface temperature of a section of the material using a traditional sensor, eg thermocouple, in order to obtain the ratio of thermal output. Temperature, however, is not necessarily directly related to moisture content.

As a general rule, the surface temperature difference between dry and damp limestone is about 0.15–0.3°C as a result of differential specific heat, when ambient temperature has risen about 5°C. Evaporative cooling is best assessed individually for each test.

Units and scale Most systems will measure in degrees Celsius and have a range to cover all likely conditions and emissivities.

Sensitivity Most modern high quality cameras have sensitivity to at least 0.1°C. Greater sensitivity is usually of little advantage; the thermal noise of the atmosphere will tend to swamp any meaningful signals below the level of about 0.075°C.

Accuracy The skill and experience of the operator in selecting the optimum conditions and optimising those conditions on site will ultimately allow the best and most accurate information on moisture to be collected and interpreted. This involves avoiding signal contamination with stray reflections, unwanted heat sources and general misinterpretation of the environment.

5.4.3 EQUIPMENT

A number of systems are commercially available, each with slightly differing characteristics. Consideration at purchase may be given to a number of different characteristics which may optimise performance for particular requirements. These will include thermal sensitivity, stability, ease and flexibility of use, spatial resolution, recording range and portability.

Some very simple cameras exist which are intended only for observation and instant analysis, without any permanent record being taken. These are not recommended for moisture detection.

Components Various typical camera set-ups are available, ranging from the hand held video camera style device (often useful for working in confined spaces or areas with difficult access) to a portable process station system (more suited to major extensive surveys). Note, however, that the small camera does not have a continuous video recorder. This would have to be a separate unit connected to the process station.

The electrical supply can be from small secondary cells (NiCad/metal hydride). For extended observations where trailing cables and mains power is inappropriate, 12 V DC lead acid batteries provide a simple solution.

Size From hand held to suitcase-sized.

Weight The video camera style system is the lightest single unit, weighing 3 kg, but a full recording station may require more equipment. The battery supplying the system may be the heaviest component.

Storage Most commercially available equipment is small enough to be carried by hand and can therefore be readily transported. Although robust in use, it is sufficiently delicate to cause concern with airport baggage handlers and it is strongly recommended that the sensors and lenses are always carried as hand baggage.

Calibration Manufacturers should carry out this task, however simple confirmation of calibration should be carried out on site at regular intervals.

5.11.4 METHOD OF OPERATION

Requirements The system is scanned over the surface in exactly the same manner as a taking a video review of the surface: radiated energy is captured by the internal sensors and converted to a scanned video image.

Samples Wherever possible, sample surfaces known to be damp or dry should be found to provide a base calibration. This confirms that the process is likely to be successful. As thermal imaging frequently relies on identifying changes of state due to the presence of moisture, eg heating or cooling cycles, the thermal output of the samples should be re-checked through the course of the survey.

Environmental The importance of considering the weather conditions and their probable effect on the subject of investigation cannot be over stressed.

Situational No particular access requirements other than a direct line of sight to the area to be tested and an appropriate viewing distance.

Human resources/ skills Generally a technician capable of operating the camera and recorder will provide the images. These will require interpretation and evaluation that may be undertaken by the specifier (with the appropriate experience) or, more commonly, by a consultant specialising in thermographic inspection. It is often advantageous to put such expertise on site to ensure that the optimum images are taken.

Initial trial testing Given the variability of UK weather conditions and the frequently narrow time windows of optimum conditions, the luxury of an initial trial as a separate task may lead to unacceptable delays in the overall project. Check calibration procedures during the course of a survey should provide an adequate proof and validation method.

Use Once the site and target has been evaluated and the most favourable conditions selected, the imager is set up at a distance from the target and images are collected of the surface in view. These may be taken as a continuous video review or detailed stills, giving the opportunity for higher resolution and thermal interference from the atmosphere to be averaged out.

Data recording Data may be collected in continuous video format, or as a series of stills. The recording may be either analogue or digital, the latter allowing for considerable post processing of data and possible enhancement of detail. When tracing moisture the thermal differences sought are often very small. Thermal noise from the air between the surface and the camera may mask these differences but they can be identified by stacking and sampling digital images. This increases data collection time but may make otherwise worthless data invaluable.

Results The images produced by a thermal camera can be as informative as they can be confusing. They need careful interpretative explanation, such that the thermal differences resulting from moisture or its consequences are distinguished from other effects. Once this is done, the inferential evidence provided by the images can be extremely helpful to the remedial planning. There may, on the other hand, be so many images that it is more sensible to require the service provider to transfer the key interpreted information to a summary drawing or plan.

Alternatively, the investigation may have been carried out to assess and provide an explanation for a particular problem. In this case the data and the thermographs may form part of a logical audit trail to the explanation of the mechanism or source of a damp problem. Again the expert service provider should give the answer to the specification in a clear and convincing manner, with sufficient data, explanation and discussion to ensure that the client has both a clear understanding of the results and confidence in the interpretation.

It is also very easy for the wrong expertise to be placed on site: rising damp and salt efflorescence produces surface damage to limestone, which is both obvious to a client and produces a strong thermal image. The client will waste money if the thermographic inspection reports only this. The thermographic inspection does itself a disservice if it misses within those strong signals the indicators of more serious and less obvious problems (see Section 5.11.5 *Case studies*).

Great attention should be paid to examining the surroundings, particularly when an anomaly has been observed. This should ensure that no unexpected physical cause of thermal variation is present: surfaces which are not normally considered reflective of visible light may be highly reflective of infrared; an operator or observer will output considerable heat, as also do tungsten lights, motors and so on.

Like many other non-destructive tests, thermographic inspection will collect a range of information on the condition of the structure as a whole. Care should be taken that this information is not lost to the overall investigation: a thermal survey may be specified to find moisture distribution in a wall as a marker for potentially damaged render, but actually locates the areas of detached render as well as damp areas. The spall data is equally valid and serves to show that the suspected process is further advanced than originally supposed. The measurement of moisture within the fabric of a building is rarely an end in itself. It is more frequently part of an overall investigation to assist in the maintenance and repair of the structure.

Digital stills may require processing software to provide further off-site analysis, plus a graphics package to display or print the results.

Specification

The cost of suitable equipment is high and purchase is not normally considered an option. Rental costs are similarly high thus it is assumed that a service provider will be sought.

The purpose of this guide is to allow the specifier to judge the technical merit of a supplier's proposals, and the accuracy with which the needs of building and problem to be investigated have been understood.

When acquiring "expert" services the specifier should invite proposals stating clearly what information is required. The specification of "a thermographic" survey leaves it up to the supplier as to whether the client receives any information on moisture! This principle should be extended to avoiding specification of the equipment or methods to be used as this will restrict the effectiveness of the survey to the limit of the specifier's understanding.

The specification should therefore clearly state the nature of the problem, the extent over which it is assumed to exist and the level of information needed. Expect the supplier to query the description of both the problem and the building in order to increase his understanding of the requirements.

5.11.5 CASE STUDIES

Leaking floor membrane

Construction New basement floor, 250 mm thick in a hydrostatic head of approximately 2 m.

Reason The membrane and stitch grouting had failed and slab joints were leaking. The test was required to find the location of failures and to confirm the success of regrouting, before laying vinyl floor (see Figure 5.11.2).

Results The centres of leakage were readily found, even when floor was damp and the technique continued to locate leaks, which caused evaporative cooling, but which could not be seen by eye.

Success Thermographs identified the exact locations of each leakage so those repairs could be carried out quickly.

Assessment This low level of leakage would have caused the floor finish to fail, although moisture was barely detectable. The technique provided a quick and efficient test on 2500 m^2 of floor.

Thermographs of floor

Before grouting, with leaking joint showing dark | **After grouting, with no evidence of thermal difference at the joint**

Figure 5.11.2 *Leaking floor membrane*

Water penetration through cladding

Construction Concrete cladding over steel frame.

Reason Water damage on interior finishes was observed together with excessive humidity in the interior. This was assumed to be caused by poor concrete or cracking of wall panels.

Results The areas of potential damp were identified below eaves, in pattern indicative of failing rainwater goods (see Figure 5.11.3).

Success Thermographs identified the exact locations of each blocked gutter, so spillage could be stopped. No further problems of moisture penetration to the interior were detected.

Assessment Thermographic inspection located the extent and position of damp in the walls allowing the correct mechanism for envelope failure to be assessed. All the rainwater goods failures spilling onto walls were found within eight hours.

Note the darker joint lines between concrete panels and very dark cool patches where spillage from guttering has occurred.

Figure 5.11.3 *Thermograph of cladding*

Water penetration through wall

Construction Stone facings of a 15th-century church wall.

Reason The interior finishes were being damaged by moisture at the base of the wall, which was initially assumed to be localised "rising damp".

Results The ingress of damp was shown to be largely "falling damp" arising from failure of the seals around the lights in the window reveals. Damp was shown to track along mortar beds and to have collected on the massive plinth stones at the base of the wall (see Figure 5.11.4).

Success Thermographic inspection corrected the initial assumptions on the source of moisture contamination. It identified the exact location of water penetration, such that timely repairs could be undertaken.

Assessment This case study demonstrates the importance in damp assessment of structures to correctly identify the source of the moisture, also of the need to review the distribution of moisture in the fabric, rather than placing undue reliance on "simple" measurement of moisture content. It clearly demonstrates how thermographic inspection can assist in this process.

Note water penetration from window reveal into mortar beds and down to base of pilaster

Figure 5.11.4 *Thermograph of aisle screen*

5.11.6 RELEVANT DOCUMENTS

Standards and legislation

None available for this type of investigative use of a technique originally designed for performing temperature measurements.

Guidance

Ballard, G S, *Non-destructive investigation of standing structures,* Technical Conservation, Research and Education Division, Historic Scotland, 1999

Hart, J M, *An introduction to infrared thermography for building surveys,* BRE, Watford, Digest IP 7/90, July 1990

5.12 Ultrasonics in the location of leaks

5.12.1 APPLICATION AND USE

Principle

The term ultrasonics refers to the broad subject of sound above the audible frequency range. This can be used to detect air paths, which may also be water paths, through a building element.

Other names

Ultrasound.

Application

The typical situation in which ultrasonics may be used is in tracing air and water leakage through vertical or sloping cladding systems. This only identifies the presence of an air path, thus positive results require further appraisal to determine their relevance to possible water leaks. It should be emphasised that not all air paths will cause water penetration so findings must be used with suitable caution.

Usage

Materials The test can be used on any solid materials but is best suited, in tracing air and water leakage, to the following:

- glass
- sealants and gaskets
- stone
- metal
- plastic and equivalent cladding materials.

Elements The most suitable elements of a building for testing using ultrasonics are:

- external walls
- sloping glazing systems.

After careful consideration of the mode of operation of the test and the form of a structure, the user may identify other individual situations in which use of the test method may be appropriate.

Use with

Ultrasonics will provide only part of the total picture necessary to identify the source and cause of any leakage problem. Other complimentary techniques will have to be selected on the basis of the particular situation. These might include hose testing, partial dismantling and borescope inspection but moisture content measurements are less likely to be required. Consideration should always be given to any available design information that may assist in the appreciation of the method by which a cladding or glazing system is intended to function.

Health and safety

There is no known health and safety risk associated with use of the battery operated equipment; the receiver is passive and the output power level of the transmitter is very low.

Costs and time

The speed of test is determined by the purpose of the testing and access limitations. Under reasonable conditions, a speed of several linear metres per minute could be possible. Note that when used on a cladding system one would normally be appraising joints and other linear features, as opposed to an area related measure.

Repeat runs with the transmitter in different locations are recommended. Slower, more precise use is recommended to examine specific leaks in detail.

Ideally two people are required to operate the system: one positioned inside the building, the other outside.

Relevance

The test method is unlikely to be of more than occasional use for detecting leaks. The most relevant application is in the assessment of cladding systems such as curtain walling, panel systems and equivalent arrangements where few, if any, air paths are supposed to remain after construction.

In the role of seeking air leaks its capabilities are far wider extending to quality control checks on glazing, doors, fire barriers etc.

Advantages

- The technique is quick, quiet and not intrusive. It does not require a supply of water and so is unlikely to cause significant disturbance to building occupants.

Prompts and pitfalls

- The principal disadvantage is that it does not positively confirm the actual route of any water and so is likely to be used only as a preliminary test tool for later supplementary appraisal by visual inspection, hose testing, partial dismantling or another equivalent technique.
- The use of ultrasonics has not developed in the building industry with the technique being very rarely applied. In consequence there is little available authoritative advice.
- If an air path were filled with water the ultrasonic sound may not transmit with sufficient power to register it as a probable leak

Hints on accuracy

- The receiver can locate the source of ultrasound quite accurately whereas the position of the transmitter has much less effect on the strength of the received signal. When examining a leak the transmitter should be placed or held as near to the known position of water ingress as possible and the receiver moved around the external face to find the inlet point, which is normally the unknown location in an investigative exercise. The receiver is always used on the "unknown" side.

5.12.2 PRINCIPLES

Property

The test method derives from other parts of the engineering industry in which the escape of gas or steam from pipework is commonly sought by use of detectors, which sense the ultrasonic noise generated by gas escape. References to this use of equivalent technology may be identified on the Internet.

In the late 1980s the technique transferred to the construction industry but it has never been widely used.

The underlying principle is that ultrasound can pass through very narrow spaces so if a small sound generator is placed on one side of a cladding system and a listening device on the other side, the cladding can be scanned to locate points at which ultrasonic noise emerges. Any such point will be one end of an air path.

Ultrasonic sound is transmitted reasonably efficiently in air but does not readily pass into and back out of solid objects. The sound readily negotiates tortuous paths. Any water leak through cladding must involve a continuous path extending between the interior and exterior of the building. This path would inevitably also be a path for air. A sound source placed at one end of the path may be detected at the other end thereby enabling an operator to locate sources of air ingress into a building. While all identified paths must be continuous air paths and hence sources of air leakage, they are not necessarily routes of water leakage so further appraisal or testing may be required.

Test basis A transmitter generates ultrasound that is emitted through a small piezzo-electric device. An equivalent sensor detects any ultrasound reaching the receiver. This is electronically converted into an audible signal, the volume of which is related to signal strength. The instrument may incorporate a visual LCD display that provides a silent measure of the strength of the incoming signal.

Measurements The ultrasonic sound, generated either by air rushing through a narrow air path in a cladding system or emitted by a transmitter and passing along air paths, is detected by the receiver, which emits audio and visual signals.

Units and scale There are no units of measurement; the equipment has no calibration. Even though larger defects may allow more transmitted sound to pass and hence a louder output to be generated, there is no direct correlation between signal amplitude and the size of the opening. This is because there are too many variables, including path length, for a meaningful and reliable comparison to be made.

Sensitivity The sensitivity of an uncalibrated device cannot be described in quantitative terms but, from a practical viewpoint, the equipment can detect gaps through what seem to the unaided eye to be closed butt joints. However, no gap is detected across a properly compressed gasket because there is no continuous air path.

Accuracy The position of apparent defects can be located to within ±5 mm if an extension tube is added to the sensor, such that the position being tested is more precisely defined.

5.12.3 EQUIPMENT

Components The main component of the test equipment is a hand held ultrasonic sound sensor. This converts ultrasonic sound to an audio signal, with an optional visual display of signal level. A headset or earphones may be added for preference or to minimise disruption. In addition a small ultrasonic sound generator is required for many building industry applications; this device is small enough to be enclosed in one hand.

A simple plastic tube added to the test equipment restricts the size of the area under detailed examination and may help to pinpoint a defect.

Size Hand held meter that would normally be battery-powered.

Weight The test equipment is lightweight, 500 g.

Storage Dry environment.

Calibration There is no calibration requirement. In use a functional check is rapidly performed.

5.12.4 METHOD OF OPERATION

The device is an ultrasonic detector (see Figure 5.12.1). The operating frequency varies between manufacturers but there is insufficient practical experience to compare the merits and drawbacks arising between the higher and lower ranges of ultrasound.

There are two general modes of operation.

1. Listening for naturally generated sound.
2. Sensing sound generated by a transmitter.

In the first mode, a single operative could undertake a survey. This type of survey would probably be conducted from within the building to locate air paths. Proving an association between air paths and the presence of any leaks is a separate operation that must form an essential supplementary exercise if water penetration is the subject under investigation. This mode of test is only effective when there is a differential air pressure between the outside and inside of the building as this would be required to cause air to "whistle" through the gap creating a noise that the sensor could detect. The test would have to be conducted on a somewhat windy day.

In the second mode of operation, two operatives are generally required: one to position and manoeuvre the transmitter, the other to operate the receiver on the other side of the barrier, normally a building envelope. The receiver should be operated on the side of the barrier on which the location of any defect is required. For a general sweep it is normally preferred to use the receiver on the interior with the transmitter outside but, to assess any individual zone in detail, the reverse arrangement may be more appropriate as it will locate the point at which water enters the system.

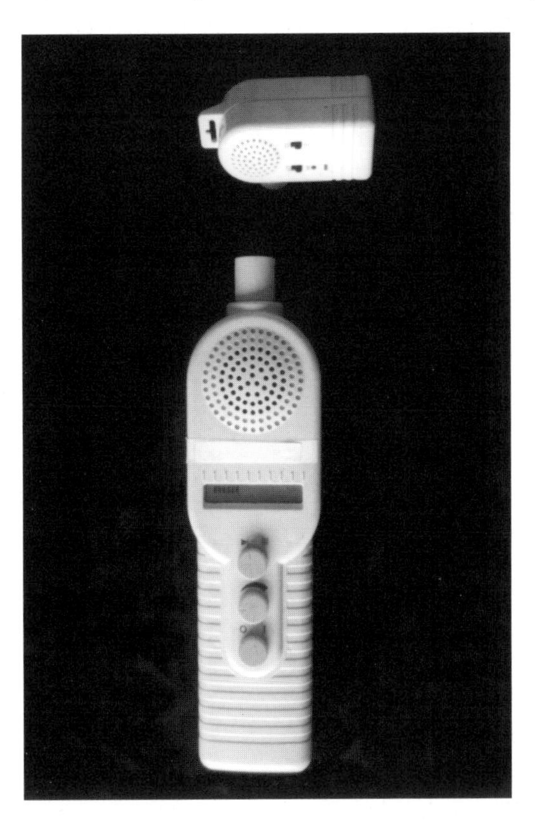

Figure 5.12.1 *Ultrasonic test equipment comprising transmitter and receiver*

Data recording The identified locations will generally be points at which air leakage occurs. However, when the transmitter and receiver are very close together and are only separated by a single layer of material, some low-level signal transmission does tend to occur through the solid materials. This is readily distinguished from true leakage paths.

Results The identified locations will generally be points at which air leakage occurs. However, when the transmitter and receiver are very close together and are only separated by a single layer of material, some low-level signal transmission does tend to occur through the solid materials. This is readily distinguished from true leakage paths.

5.4.5 CASE STUDY

Ultrasonics has rarely been used for this type of application in the UK.

5.4.6 RELEVANT DOCUMENTS

Standards and legislation There are no BS or equivalent standards applicable to the use of ultrasonics in leak detection within the building industry.

Guidance No guidance documents have been identified.

6 Automated long-term monitoring

6.1 APPLICATION AND USE

Principle The principle of automated long-term monitoring, related to moisture in building elements, is to measure and record physical conditions over time using electronic sensors linked to a data logging system, and to process and present the data so that it can be used to effectively control these conditions.

Other names Monitoring systems that may appear to carry out similar functions to those described above are often used in buildings. It is therefore important to understand the uses and limitations of such equipment, as detailed in Table 6.1.

Table 6.1 *Building monitoring systems*

System	Application
Building management systems	Most large heating and ventilation systems include an electronic building management system (BMS), usually based on a specialist PC or data logger. They monitor the air conditions within the building, which are only indirectly related to the moisture in the building elements.
Leak detection systems	Leak detection systems are sometimes installed on plumbing and water pipes in building structures in areas particularly vulnerable to water damage, such as computer rooms or libraries. These give a local alarm at a base unit if water penetration is detected but are notoriously subject to "false alarms" from, for example, condensation. However leak detection sensors may provide a useful function as part of any automated long-term monitoring system.
Museum monitoring systems	A number of specialist data loggers have been configured for monitoring the condition of valuable artefacts in museums or collections. These are able to monitor rh, temperature and light, but are not generally configured to monitor moisture in materials.

Application Moisture in buildings often varies significantly with time, eg daily due to occupancy and the natural diurnal cycle and annually due to seasonal changes. Section 2 *Understanding moisture* provides guidance on the main parameters relating to moisture problems. For many of these, a long time period is required for significant moisture changes to occur, eg several years for the moisture contents of building elements to settle down after construction or refurbishment. Automated long-term monitoring of physical conditions related to moisture in building elements may therefore be cost-effectively applied in a wide range of circumstances, as shown below.

1. To aid diagnosis of the causes of high moisture levels.
2. To determine the effectiveness of moisture control measures, for example:
   - after fire or flood
   - during accelerated drying
   - following remedial works.
3. To monitor the moisture levels and behaviour in buildings, elements and materials.
4. To allow cost-effective facilities management by enabling measurements to be taken:
   - remotely with control to local and remote alarm systems
   - in inaccessible parts of the structure
   - without disturbing occupancy.

Automated long-term monitoring of moisture in building elements has proven beneficial:

- in the conservation and refurbishment of historic buildings
- in the management and maintenance of large corporate or government buildings, buildings in multiple occupancy and buildings in intermittent occupancy, such as churches or holiday homes
- in the diagnosis of moisture-related problems in buildings and in the testing of proposed remedial materials or detailing.

For the future, especially in conjunction with telecommunications, automated long-term monitoring systems will be of increasing benefit:

- in buildings generally to make facility management and maintenance more cost-effective
- in new buildings and system building elements, both for quality control and maintenance purposes: for example, they will be increasingly used to monitor the drying of concrete during construction.

Usage

Materials and elements

Automated long-term monitoring is applicable to all buildings, elements and materials. Example uses include:

- the masonry of external walls
- timber components, especially those in contact with external walls
- the materials making up roof decks or wall to floor junctions
- the elements making up roof drainage or plumbing systems in vulnerable areas.

Test methods

Most physical characteristics of buildings, elements and materials can be electronically and hence automatically measured, and a large number of different types of electronic sensors are available. The principles of the techniques specific to moisture measurement are described in Section 5 *Test methods* and their application to automated long-term monitoring identified in Table 6.2, together with other relevant parameters.

Table 6.2 *Test measurements used in automated long-term monitoring systems*

Test measurement	Comment
Relative humidity (environment/ condition)	Section 5.5 gives details of a range of rh sensors. Timber plugs and electronic rh probes are suitable for automated long-term monitoring.
Resistivity (moisture content)	Section 5.4 describes the electrical resistance meter; a purpose built meter using a two-probe technique measuring resistivity and having moisture "scale" calibrated for specific materials. Two probe or four probe (Wenner) methods are used with automated long-term monitoring systems and provide a direct but generally only a comparative indication of material moisture condition. The rh measurement from timber plugs, Section 5.5, is generally based on a resistivity measurement of a "standard" material with known moisture related properties.
Leak detection (liquid water)	Section 5.3 describes the electrical earth leakage technique for the location of leaks. Leak detection tapes and systems are designed using a similar technique but used to detect moisture ingress but not necessarily the source. Sensors based on capacitance, Section 5.2, and optical fibres, not covered in this guide, may be considered in specific applications.
Temperature (related factor)	The selection of temperature sensors has not been covered in this guide. Thermo-couples, thermistors and resistance sensors are commonly used with automated long-term monitoring systems. Temperature is an important factor in the interpretation of moisture measurement data.
Strain and movement (related factors)	The selection of strain and movement sensors has not been covered in this guide. Electrical resistance and vibrating wire strain gauges and / or electrical resistance and linear inductive movement gauges are commonly used with automated long-term monitoring systems. Moisture is one of factors that contribute to strain and movement changes; other factors include temperature and loading.
Chemical and biological (related factors)	The selection of chemical and biological sensors has not been covered in this guide. Dedicated sensors are available to monitor the concentration of key molecules in the air that would indicate initiation of decay by one of these causes due to the ingress of moisture.
Operational and environmental (related factors)	Operational and environmental effects have been discussed in Section 2 but not related sensors. Internal and external ambient temperature and rh are important factors. Consideration should be given to the installation of a weather station (wind speed and direction, rainfall and solar power). Establishing links to building control systems may also prove a useful source of building performance data (heating, ventilation, lighting, usage and occupancy).

Use with

It is usually necessary to carry out a condition survey of the building elements prior to the design and installation of any system of monitoring. This will help to correctly specify the problem to be addressed and to identify the appropriate sensors and critical locations. The basis for the selection of the appropriate test methods(s) to be used in the survey is given in Section 4 *Selecting the appropriate test method* and these methods are described in Section 5 *Test methods*.

Health and safety

The installation of a building monitoring system has few health and safety issues that are not covered in the general section, Section 3.5 *Health and safety*.

Electrical wiring to the sensors should be low voltage and therefore only represents a potential hazard in the case of fire. If this is a special concern, low halogen emission wiring should be specified.

Costs and time

The factors that can affect the cost and turn-round time of installation and operation of automated long-term monitoring systems that need to be taken into consideration are:

- unit cost, durability, maintenance costs and availability of:
  - sensors
  - monitoring system base unit
  - cabling or telemetric link between sensors and base unit
- ease of installation of:
  - sensors
  - cabling or telemetric links
- ease of programming and setting up the base unit
- costs of providing a PC and associated hardware for both local and remote access to system and data; there is the need to check compatibility and availability if considering sharing this resource with other applications
- cost and ease of use of analytical and data handling software (check licensing agreements)
- cost, availability and ease of use of remote telecommunications equipment/packages
- cost of personnel to manage the system and collect and interpret data throughout the operation period.

Relevance

Automated long-term monitoring systems are designed to enable the moisture content of building materials to be remotely monitored to allow building failures or defects to be detected and diagnosed at an early stage. They are particularly useful for checking the moisture content of inaccessible moisture sensitive materials in roof spaces, behind decorative finishes and in walls. In many cases such monitoring systems can increase the efficiency of inspection and maintenance programmes and hence reduce building life-cycle costs. The reduction of potential health and safety risks associated with access to the building elements may be a major factor in deciding to install a system.

Advantages

- Measurements over time allow diagnosis of defects, assessment of performance of elements under field conditions and hence better assessment and management of risk.
- Automatic measurement allows multiple repeat measurements and hence minimises staff time in taking such measurements.
- Automatic recording allows cost-effective management and analysis of data by minimising the staff time required for data handling. This also minimises the opportunity for errors to be introduced during recording or data handling.
- Remote measurement minimises:
  - the staff time and costs of access for repeat measurements
  - disruption of occupancy to obtain measurements
  - possible health and safety risks to staff in carrying out measurements.
- The use of remote operation via the telephone system allows:
  - projects to be managed from any location with telephone access
  - minimises the cost of accessing the system and data
  - remote facilities management to be alerted of alarms or changes via telephone or e-mail.
- The system can also be used to monitor other parameters/conditions providing a comprehensive linked database to facilitate effective building management.

Prompts and pitfalls

- Adequate provision must be made for monitoring, maintenance and repair during the intended operating life of the system. This should avoid the potential problems of waste of resources such as expensive hardware that is ineffectively used by generating data that no one can interpret.
- Provision should be for adequate and reliable power supply that provides backup in the event of a mains power supply failure. Similarly, provision should be made for the "in-time" replacement of batteries for battery-powered units.
- The sensor locations and the position of the cables must be accurately recorded.
- A large amount of data is generated. This must be analysed and correctly interpreted so that useful information can be passed to the appropriate decision make at the right time. Remote operation via telecommunication can facilitate data recording and analysis.

Hints on accuracy

- Clearly define the problem and plan the system carefully to ensure that:
  - the right type and number of sensor are used
  - the sensors are located at the critical points in the building structure or material
  - appropriate recording intervals are used
  - "robust" equipment and methods are used to collect and process data
  - experienced staff is used to carry out the installation and management of the system and data
  - there is a clear plan for monitoring analysis, maintenance and information dissemination.
- Use a large number of low-cost sensors rather than a few high-cost sensors in order to allow "pattern recognition" and "redundancy".
- Always monitor ambient conditions that may influence the parameters under investigation for example internal and external temperature and rh.

6.2 PRINCIPLES

Automatic monitoring of moisture in building elements relies on the use of remotely read electronic sensors. Each system should be specially designed or "customised" for the specific application.

Property

The most widely used and cost-effective monitoring systems measure the physical parameters associated with moisture by measuring their effect on electrical resistance. This requires equipment capable of measuring across an impedance range from ohms (Ω) to hundreds of mega ohms, as indicated in Table 6.3.

Table 6.3 *Resistance measurement sensors used for moisture monitoring systems*

Sensor	Parameter	Typical impedance
Electrodes in timber, either in the parent material or rh sensor	resistivity	MΩ
Electrodes in concrete (or other materials)	resistivity	KΩ
Thermistor	temperature	KΩ
Linear resistance track	movement	KΩ
Electrical resistance gauge	strain	Ω

Other sensor systems use, for example, electrical capacitance, light. However, these require additional and generally more complex circuitry, which increases the cost. They tend to be used in applications where it is inappropriate to use the more conventional resistance measurement techniques, for example where there is a requirement for higher accuracy, or the use of non-metallic (optical fibre) sensors that would not be influenced by electrical noise.

Measurements

The most cost-effective systems mainly use very small, low cost sensors to monitor material moisture content in building elements. These measure the variations in resistance of microcells made of materials of known characteristics to determine their moisture content and hence the "available moisture content" of the material in which they are embedded (see timber plugs in Section 5.5 *Humidity sensors*).

Units and scale

The sensor data are generally converted to an appropriate unit by the software using appropriate calibration coefficients or "look-up" tables. In most cases, comparative readings are produced as it is the change in reading over space and time that is most significant.

Sensitivity

The sensitivity of an automated long-term monitoring system is mainly limited by the sensitivity of sensors deployed. The least sensitive data logging systems use an analogue (sensor output) to digital (data value) conversion (A/D) with 256 incremental changes across their measurement range (eight-bit A/D conversion), which, if matched to the sensor output range, will provide sufficient sensitivity for the majority of applications.

Accuracy

The accuracy of a monitoring system is dependent on the selection of appropriate sensors, data logging system and the care taken when these are installed. Other factors potentially reducing the accuracy are condensation, chemical contamination, corrosion, physical damage and calibration "drift" of both sensors and logging system. For most practical purposes, the accuracy of reading from any sensor is at best ±0.5 per cent moisture content by dry weight in wood.

6.3 EQUIPMENT

Components

Automated long-term monitoring systems components include sensors, connecting wires, communication nodes or transmitters and base unit.

Many proprietary data loggers are used in other areas of science and industry. Care must be exercised in their selection, as many are limited as to the type and number of types of sensor they can manage. The cost of the large amount of wiring that may be required from each sensor to a central data logger can also limit their usefulness. To help overcome this problem some data loggers can be linked as outstations around groups of sensors and communicate the data to a central system. Other systems use radio telemetry to transmit data between the sensors and base unit. This can reduce the problems of wiring but increases the cost per sensor and brings the problem of radio transmission within the building.

Size

Sensors/probes vary from small pin-sized electrodes to proprietary probes that are dependent on the size of the protective housings to the sensor element and associated electronics (**transmitter**).

Nodes vary from the size a matchbox to that of a briefcase. Radio telemetric transmitter nodes are about the size of a mobile telephone.

Cabling may be as small as 3 mm diameter, mainly for communication. Analogue systems may require the use of screened multi-core cable typically up to 10 mm in diameter. The cable diameter and configuration should be selected taking into consideration the use, electrical environment and length.

Base units vary from briefcase to suitcase in size dependent on the number of sensors they support. Most systems base units will also require a PC for data analysis. Modular systems are now available for installation within a desktop PC or in an industrial PC (210 mm × 160 mm flat panel) with touch screen and graphics.

Calibration

Sensor calibration should be traceable to National Institute of Standards and Technology (NIST) reference measurements under United Kingdom Accreditation Service (UKAS), or similar, certification.

Sensors should be calibrated before installation. Sensors requiring recalibration in less than the planned monitoring period should not be used unless this can be done remotely or they can be cost-effectivively re-placed.

6.4 METHOD OF OPERATION

Requirements

Samples Sensors should be installed in small groups to allow some redundancy and aid interpretation. Usually the minimum number of moisture sensors required to give a meaningful result is four.

Environmental The sensor, associated electronics, connectors and cabling must be adequately protected against weather and vandalism.

Situational Adequate provision should be made to ensure that the base unit and associated equipment is adequately secured to against accidental damage and tampering. The enclosure should also have a controlled, condensation-free environment.

It will generally be necessary to record ambient conditions inside and outside the building envelope. The system should therefore allow interfacing with a standard automatic weather station.

Human resources/ skills If an automated long-term monitoring system is to be cost effective a high level of expertise is required of the person responsible for the design, installation, operation and in the analysis of the data. Non-expert staff can monitor established systems provided appropriate software has been programmed to give alarms and generate easily followed instructions for actions.

Initial trial testing No initial trial testing of an automated long-term monitoring system is required. In practice, it has been found that it generally takes a few days to establish and programme a system so that it is performing satisfactorily. Modification of the system may be required with experience and when moisture patterns in the structure have settled down.

Use

The most cost-effective way to set up automated long-term monitoring equipment is to position the sensors first. The location of the base unit will be decided with regard to the availability of suitable power supply, communications facilities and the convenience of the intended operators. Cabling or other links are then laid with regard to ease of access for installation and maintenance.

The time interval between measurements will vary with the expected rate of change for the conditions being investigated. This may vary from 30 seconds for sensors monitoring leaks or water ingress into vulnerable areas to daily for the drying down of structures. To prevent data overload, monitoring systems should be designed to check sensors measurements at the required short time intervals, but only store data when changes occur or at maximum time intervals, typically four or six hours. The system can be designed to use a more frequent sampling rate when a particular change in sensor reading is detected.

Data recording

Data should be recorded by the use of industry standard hardware and software to allow easy interfacing with other systems and software.

Results

The key to interpretation of data is to make appropriate comparisons with similar situations in space and time. The most important aspect of use is the dissemination of information to the right people at the right time in a form that can be easily understood. This generally requires that the analyst must have ready access to the data and that the resulting information should be automatically disseminated to all interested parties. Graphical representations of the data on building plans using IT graphic software and telecommunications links have proved very helpful to meet this objective.

Specification

The system should be able to interface to the wide range of types of electronic sensors required for automated long-term monitoring of moisture related parameters in building elements. In particular the system should be able to monitor the moisture content of structural timber, masonry or other materials; temperature; relative humidity; other moisture related parameters such as movement and to detect leaks.

The system should be able monitor sensors that will generally be widely dispersed throughout the building. It should be able to collate the readings from these sensors at one central location, and to manage to the volume of data generated.

The system should provide an alarm at the central location when any of the sensors give readings outside an upper or lower limit. The alarm should be capable of being independently set and variable for each sensor. All alarms should be recorded together with sensor data for future analysis and the system designed to minimise "false alarms". The system should be able to automatically generate an alarm remotely if required.

The system should be able to monitor readings from all sensors at selected intervals, from seconds, if necessary, to hours and include computer software to enable a competent building manager to analyse and act on the data recorded.

The system should be capable of being managed and read remotely to allow holistic analysis of problems by externally located experts or remote management teams. Communication will usually be via a modem and the public telephone system.

The system should be capable of future expansion to cover adjacent areas and to interface with other building management systems.

The installers of the system should provide expert consultants and technicians for the positioning and installation of sensors and continuing maintenance and management of the system.

6.5 CASE STUDIES

Brighton Pavilion

The system, designed more than ten years ago, used paired stainless-steel screws wired back to a junction box, with readings taken using a standard resistance meter. This technique was subsequently used successfully at the Mansion House, London, as well as in many more modest properties.

Research

As part of a long-term trial for Building Research Establishment the Timber Research and Development Association is monitoring nearly 1000 window frames using a single system that can monitor up to 4000 sensors.

Remote monitoring

Buildings currently being monitored remotely include Windsor Castle, the Royal Courts of Justice, York Crown Court, the Mansion House, Horse Guards and several private palaces or homes, and some small churches including Dragar Kirk in Denmark.

Special projects

Special projects can also be monitored for diagnostic or experimental purposes. The Monument in the City of London was remotely monitored for two years for the Corporation of London. More recently, the Empress Place Museum Buildings in Singapore are being remotely monitored from the United Kingdom for the Singapore Government. As part of an international collaborative project a timber church was monitored in Romania.

Flood and fire

Remotely operated automatic monitoring systems have been successfully used to control drying and proper refurbishment after fire and flood. These include Hampton Court and, more recently, at Windsor Castle where the monitoring systems were instrumental in allowing refurbishment ahead of programme and under budget.

6.6 RELEVANT DOCUMENTS

Standards and legislation

There are no current or proposed BS/CEN standards covering automatic building monitoring systems, apart from general standards covering the installation of electrical systems in buildings.

Guidance

No guidance documents or directives from BRE or other similar government organisation covering automated monitoring systems have been identified.

Key reading

Hutton, G, *Building maintenance: the H&R 'curator' and building monitoring systems,* Museum Management and Curatorship, Butterworths, Vol 14, No 1, pp 92–105, 1995

7 References and bibliography

7.1 REFERENCES

Ahmet, K, Dai, G, Jazayeri, S and Tomlin, R, *Tests on the use of timber moisture meters*, Journal of the Association of Building Engineers, **72**(6), pp 10–13, 1997

Ahmet, K, Jazayeri, S and Hall, G, *Standardisation of conductance-type timber moisture meters*, Proceedings of the 7th international conference on the durability of building materials and components, Stockholm, edited by C Sjostrom, E & F N Spon, London, pp 673–682, 1996, ISBN 0 419 20690 6

Ahmet, K et al, *The moisture content of internal timber*, Journal of the Association of Building Engineers, **70**(2), pp 18–20, 1995

Ahmet, K et al, *The moisture content of internal timber: 2*, Journal of the Association of Building Engineers, **71**(3), pp 10–14, 1996

American Standards for Testing and Materials, *Direct moisture content measurement of wood and wood-base materials, Test method for*, ASTM, West Conshohocken, PA, ASTM D 4442: 1992

American Standards for Testing and Materials, *Use and calibration of hand held moisture meters, Test method for*, ASTM, West Conshohocken, PA, ASTM D 4444: 1992

Annan, A P, *Ground penetrating radar workshop notes*, Sensors & Software, private communication (available from the manufacturer)

Assenheim, J G, *Moisture measurement in the concrete industry*, Concrete Plant and Production, September/October 1993

Ballard, G S, *Non-destructive investigation of standing structures,* Technical Conservation, Research and Education Division, Historic Scotland, 1999

British Standards Institution, *Code of practice for flooring of timber, timber products and wood based panel products*, BSI, London, BS 8201: 1987, (ISO 631 NEQ; ISO 1072 NEQ)

British Standards Institution, *Code of practice for installation of chemical damp-proof courses*, BSI, London, BS 6576: 1985

British Standards Institution, *Code of practice for installation of resilient floor coverings*, BSI, London, BS 8203: 1996

British Standards Institution, *Code of practice for installation of textile floor coverings*, BSI, London, BS 5325: 1996

British Standards Institution, *Guide to building maintenance management*, BSI, London, BS 8210: 1986

British Standards Institution, *Performance requirements for electrically-heated laboratory drying ovens*, BSI, London, BS 2648: 1955

British Standards Institution, *Performance requirements for electrically-heated laboratory drying ovens,* BSI, London, BS 2648:1955

British Standards Institution, *Quality systems. Model for quality assurance in production, installation and servicing*, BSI, London, BS: EN: ISO 9002:1994

British Standards Institution, *Specification for the performance of damp-proof courses installed to prevent rising damp*, BSI, London, DD 205: 1991

British Standards Institution, *Specification for timber*, BSI, London, BS 1186: Part 1: 1991

British Standards Institution, *Specification of aggregates from natural sources for concrete*, BSI, London, BS 882: 1983

British Wood Preserving and Damp-proofing Association, *Code of practice for installation of chemical damp-proof courses*, BWPDA, London, 1995

Bruker, *An introduction to analytical applications of low resolution NMR*, Bruker minispec application note 3, 1994

Building Research Establishment, *Rising damp in walls: diagnosis and treatment*, BRE, Watford, Digest 245, minor revisions ed. 1986, reprinted 1989

Cheetham, D W and Howard, C A, *Translating research into practice. Wall dampness diagnosis – let's get it right*, Building Engineer, February 1999

Coleman, G R, *Use of electrical moisture meters*, Building Engineer, 1997

Comité Européen de Normalisation, *Round and sawn timber: method of measurement of moisture content*, Office for Official Publications of the European Communities, Luxembourg, CEN175-13.01: March 1995

Concrete Bridge Development Group, *Testing and monitoring the durability of concrete structures,* CBDG, Technical Guide No 2, to be published

Concrete Society, *Guidance on radar testing of concrete structures*, Concrete Society, Slough, Technical Report No 48, 1997

Deutsches Institut für Normung, *Hot water floor heating systems design and construction*, Beuth Verlag GmbH, Berlin, DIN 4725: Part 4: 1992

Forde, M C, McCavitt, N, Binda, L and Colla, C, *Identification of moisture capillarity in masonry using digital impulse radar*, Structural Faults & Repairs, 1996

Hart, J M, *An introduction to infrared thermography for building surveys,* BRE, Watford, Digest IP 7/90, July 1990

Her Majesty's Stationary Office, *Construction (Design and Management) Regulations 1994 ,* HMSO, London, Statutory Instrument No. 3140, Health and Safety, 1994

Her Majesty's Stationary Office, *Environmental Protection Act 1990,* HMSO, London, Chapter 43, November 1990

Her Majesty's Stationary Office, *Health and Safety at Work etc Act 1974,* HMSO, London, Chapter 37, 1974

Her Majesty's Stationary Office, *Radio Substances Act*, HMSO, London, Chapter 12, 1993

Her Majesty's Stationary Office, *The Control of Substances Hazardous to Health Regulations 1994,* HMSO, London, Statutory Instrument No 3246, Health and Safety, 1994

Her Majesty's Stationary Office, *The Ionising Radiation Regulations*, HMSO, London, Statutory Instrument No 1333, Health and Safety, 1985 (due for update January 2000)

Her Majesty's Stationary Office, *The Manual Handling Operations Regulations 1992,* HMSO, London, Statutory Instrument No 2793, Health and Safety, 1992

Her Majesty's Stationary Office, *The Personal Protective Equipment at Work Regulations 1992,* HMSO, London, Statutory Instrument No 2966, Health and Safety, 1992

Her Majesty's Stationary Office, *The Radioactive Material (Road Transport) (Great Britain) Regulations*, HMSO, London, Statutory Instrument No 1350, Health and Safety, 1996

Her Majesty's Stationary Office, *Use and limitations of ground penetrating radar for pavement assessment,* Highways Agency, HMSO, London, Design manual for roads and bridges, Volume 7, Section 3, Part 4, HA 72/94, 1994

Howard, C A, *An evaluation of the techniques employed to diagnose rising ground moisture in walls*, Liverpool Polytechnic, MPhil thesis, 1986

Hutton, G, *Building maintenance: the H&R "curator" and building monitoring systems,* Museum Management and Curatorship, Butterworths, Vol 14, No 1, pp 92–105, 1995

James, W, *The interaction of electrode design and moisture gradients in dielectric measurements on wood,* Wood and Fiber Science, **18**(2), pp 264–275, 1986

Jazayeri, S and Ahmet, K, *Moisture gradient studies in timber by measurement of dielectric parameters,* Proceedings of the 3rd International Symposium in Moisture and Humidity (Volume 2), National Physical Laboratory, pp 179–186, 1998

Jensen, V, *Use of wooden plugs to determine relative humidity in concrete*, in Norwegian. English paper in preparation for 11th International Conference Alkali-aggregate reaction in concrete, Quebec, June 2000

Johnasson, C H and Persson, G, *Moisture absorption curves for building materials* National Research Council of Canada, Technical translation 747, Ottawa, 1998

Krus, M, *Moisture movement and transport coefficients of porous mineral building materials – new measuring techniques*, University of Stuttgart PhD thesis, 1995

Matthews, S L, *Application of subsurface radar as an investigative technique*, CRC Ltd, London, BRE Report BR340, 1998

McDonald, P and Strange, J, *Magnetic resonance and porous materials*, Physics World Vol. 11, pp 29–34, July 1998

National Building Specifications, *Electronic roof integrity test*, NBS Ltd, London, J42 pp 22–23, 1995

National Physical Laboratory and The Institute of Measurement and Control, *Guide to the measurement of humidity*, InstMC, London, 1996

Nilson, L, *Hygroscopic moisture in concrete – Drying, measurements and related material properties*, University of Lund, Gothenburg, Sweden, Report TVBM-1003, 1980

Oliver, A, *Dampness in buildings* (second edition), Blackwell Scientific, 1997

Parrett, M, *Managing disrepair in local authority housing: the misdiagnosis of rising damp using electronic moisture meters*, Professional paper 97009, Lewisham Council, 1997

Parrott, L J, *Factors influencing relative humidity in concrete,* Magazine of Concrete Research Vol 43, No 154, pp 45–52, March 1991

Parrott, L J, *Moisture profiles in drying concrete*, Advances in Cement Research Vol 1, No 3, pp164–170, July 1988

Pel, L, *Moisture transport in porous building materials*, Eindhoven University PhD thesis, 1995

Roberts, K, *The electrical earth leakage technique for locating holes in roof membranes*, Proceedings of the Fourth International Symposium on Roofing Technology, National Institute of Standards and Testing, Washington, USA, September 1997

Roofing Cladding and Insulation, *Flat roof leak detection: using portable electrical conductance technique*, RCI, London, Technical Note No 35, p 27, January/February 1994

Roofing Industry Educational Inst., *Roof moisture surveys current state of technologies*, RIEI. Englewood Colorado USA, 1989

Skaar, C, *Wood-water relations*, Springer-Verlag, New York, 209, 1988

Smith, J L, *Experience in developing an expert system,* Frameworks for defects diagnosis, Building Research & Practice, Vol 14, No 2, pp 85–87, 1987

Structural Studies & Design, *Notes on application of wooden plugs for the determination of in-situ rh and moisture in concrete and other porous materials*, trade literature

Timber Research and Development Association, *Moisture meters for wood*, wood information, TRADA, High Wycombe, Section 4, Sheet 18, revised October 1991

Timber Research and Development Association, *Moisture in timber*, wood information, TRADA, High Wycombe, Section 4, Sheet 14, April 1999

Torgovnikov, G I, *Dielectric properties of wood and wood-based materials* Springer-Verlag, New York, 1993

Trechsel, H T, *Moisture control in buildings*, ASTM, West Conshohocken, PA, ASTM Manual series MNL 18, 1994. ASTM code pcn 28-0180094-10

Troxler International, *Roof moisture gauge*, Instruction Manual, trade literature

Wood, J G M, *Durability design: form, detailing and materials*, Building the future, pp 23–32, Garas, F K (ed), E & FN Spon, London, 1994

Wood, J G M, *Methods for the control of active corrosion in concrete*, 1st International Conference Deterioration and repair of reinforced concrete in the Arabian Gulf, Bahrain Society of Engineers and CIRIA, October 1985

Wood, J G M, Nixon, P J and Livesey, P, *Relating ASR structural damage to concrete composition and environment*, pp 450–457, Shayan, A (ed), Proceedings 10th International Conference Alkali-aggregate reaction in concrete, Melbourne, 1996

7.2 BIBLIOGRAPHY

ANON, *Highlighting heat loss from the air,* Energy in Buildings & Industry, pp 10–11, March 1996

ANON, *Leak preview,* Building, pp 62–63, 6 Nov 1992

ANON, *New uses for impedance spectroscopy,* Research Focus No 28, 28 February 1997

ANON, *Rising damp? No such thing,* Independent on Sunday, p 23, 23 November 1997

ANON, *The Autoclam,* CNS Electronics Ltd, trade literature

ANON, *Watching the detectors,* Building, pp 50–51, 27 Jan 1995

Basheer, P A M and Long, A E, *In-situ monitoring techniques at Queen's University of Belfast,* Private communication of paper presented at BRE, May 1996

Basheer, P A M, Long, A E and Mongomery, *A comparison between the Autoclam permeability system and the initial surface absorption test (ISAT),* Structural Faults conference, Edinburgh, pp 71–77

Basheer, P A M, Long, A E and Mongomery, *The Autoclam –- a new test for permeability,* Concrete, pp 27–28, July/August 1994

Bell, S A, *A study of the requirement for standards for the measurement of moisture in solid materials,* National Physical Laboratory, Report CMAN 20, June 1998

British Standards Institution, *Code of practice for flooring of timber, timber products and wood based panel products,* BSI, London, BS 8201: 1987 (formerly CP201)

British Wood Preserving and Damp-proofing Association, *The use of moisture meters to establish the presence of rising damp,* BWPDA, London, DP1: November 1993

British Wood Preserving Association, *Moisture content determination,* BWPA Manual, BWPA, London: Section 1, Part 9.1.3, 1986

Building Research Establishment, *Condensation in roofs,* BRE, Watford. Digest 180, 1986, ISBN 0 85125 219 2

Building Research Establishment, *Design of timber floors to prevent decay,* BRE, Watford. Digest 364, 1991

Building Research Establishment, *Dry rot: its recognition and control,* BRE, Watford, Digest 299, 1985, ISBN 0 85125 348 2

Building Research Establishment, *Drying out buildings,* BRE, Watford, Digest 163, 1974, ISBN 0 11 721606 2

Building Research Establishment, *House inspection for dampness: a first step to remedial treatment for wood rot,* BRE, Watford, IP 19/88, 1988

Building Research Establishment, *Interstitial condensation and fabric degradation,* BRE, Watford, Digest 369, 1992, ISBN 0 85125 519 1

Building Research Establishment, *Painting exterior wood,* BRE, Watford, Digest 422, 1997 (supersedes BRE Digest 354, 1990)

Building Research Establishment, *Painting walls. Part 2:Failures and remedies,* BRE, Watford, Digest 198, 1977 reprinted 1984, ISBN 0 11 724904 X

Building Research Establishment, *Timbers: their natural durability and resistance to preservative treatment,* BRE, Watford, Digest 296, 1985

Building Research Establishment, *Wet rots: recognition and control,* BRE, Watford. Digest 345, 1989, ISBN 0 85125 389 X

Building Research Establishment, *Why do buildings crack?,* BRE, Watford. Digest 361, May 1991, ISBN 0 85125 476 4

Coleman, G R, *A guide to the use of electrical moisture meters,* Remedial Technical Services, trade literature

Coleman, G R, *Dampness following the insertion of a remedial damp-proof course,* Structural Survey, pp 3–9, Summer 1992

Coleman, G R, *Evaluating the performance of chemical injection damp-proof courses,* Building Engineer, April 1997, p 4, Remedial Technical Services, trade literature

Construction Research Communications, *Diagnosing the causes of dampness,* CRC Ltd, London, BRE Good Repair Guide 5, January 1997, ISBN 1 86081 115 9

Construction Research Communications, *Flat roofs: assessing bitumen felt and mastic asphalt roofs for repair,* CRC Ltd, London, BRE Good Repair Guide 16 Part 1, May 1998, ISBN 1 86081 220 1

Construction Research Communications, *Treating rain penetration in houses,* CRC Ltd, London, BRE Good Repair Guide 8, April 1997, ISBN 1 86081 135 3

Construction Research Communications, *Treating rising damp in houses,* CRC Ltd, London, BRE Good Repair Guide 6, February 1997, ISBN 1 86081 126 4

Endean, K F, *Looking for leaks,* Building, Roofing supplement, p 38, 22 January 1998

Endean, K F, *Damp meters: other electrical systems,* Investigating rainwater penetration of modern buildings, Gower, pp 20–22, 1995

Fishburn, D C, *Roof thermography: Detection of subsurface roof moisture,* IRIE, pp A57–A60, 1978

Flanders, S and Yankielun, N, *Two new roof moisture sensor technologies,* 4th Symposium on Roofing Technology, Paper 49, USA, 1997

Hedenblad, G, *Drying of construction water in concrete: drying times and moisture measurement,* Swedish Council of Building Research, 1997

Henriksen, C F, *In-situ monitoring of concrete structures*, pp 165–178, private communication of paper prepared by RH&H Consult, Copenhagen, Denmark

Her Majesty's Stationary Office, *Defects in buildings,* HMSO, London, 2nd edition, 1989

Hewlett, P, Hunter, G and Jones, R, *Bridging the gaps,* Chemistry in Britain, 1999

Hooker, J, *More on thermographic flat roof surveys,* RCI, London, p 31, October 1997

Howell, J, *Moisture measurement in masonry: limitations;. hand-held metering.. & implications .. treatment of dampness ..,* 7th North American Masonry Conference, pp 773–784, June 1996

James, W L, *Electric moisture meters for wood,* US Dept of Agriculture, Forest Service, Forest Products Laboratory, 1988

Kemmsies, M, *Comparative testing of Wagner L612, electrical resistance meters, and the oven-dry determination of wood moisture content on Norway spruce and Scots pine,* SP- Swedish National Testing and Research Institute, Wood Structures and Materials, Report 97B2, 1983

Laffan, R, *Water, water everywhere...,* Concrete Engineering international, pp 18–24, May/June 1998

Nilsson, L, *Methods of measuring moisture penetration into concrete submerged in seawater,* pp 199–208, private communication of paper prepared by Chalmers University, Sweden, 1996

Oxley, R, *Damp and timber treatment,* Royal Institution of Chartered Surveyors, Do's and Don'ts Guide 3, November 1997

Oxley, T A and Gobert, E G, *Moisture meters,* Dampness in Buildings, Diagnosis, Treatments, Instruments, Butterworths, pp 26–32, 1981

Parrott, L J, *A review of methods to determine the moisture conditions in concrete (87 references),* British Cement Association (BCA) publication C/7, December 1990

Peskett, M, *Infra-red defect detection,* RCI, London, pp 23–24, February/March 1993

Pye, P W and Harrison, H W, *Floors and flooring: performance, diagnosis, maintenance, repair and the avoidance of defects,* CRC Ltd, London, BRE Building Elements, Chapter 1.4, pp 25–31, 1997, ISBN 1 86081 173 6

Ranta-aho, T, *Moisture in concrete and how to measure it,* Vaisala News, 145, pp 16–17, 1997

Roberts, K, *Report from the 9th Congress of the IWA,* RCI, London, p 32, December 1995/January 1996

Roofing Cladding and Insulation, *Flat roof leak detection: electrical capacitance technique,* RCI, London, Technical Note No 40, p 15, September 1994

Roofing Cladding and Insulation, *Flat roof leak detection: using portable electrical conductance technique,* RCI, London, Technical Note No 35, p 27, January/February 1994

Roofing Cladding and Insulation, *Infra-red thermography,* RCI, London, Technical Note No 34, p 27, November/December 1993

Saadatmanesh, H and Ehsani, R, *Non-destructive evaluation of concrete and wood properties using NMR,* Insight Vol 39 No 2, pp 75–81, February 1997

Smith, Williamson and Cudby, *Investigation of methods for the measurement of dryness in cementitious subfloors,* Private communication of paper prepared by University of East Anglia and Contract Flooring Association

Stanley, C and Balendran, R V, *NDT of the external surfaces of concrete buildings and structures in Hong Kong using infra-red thermography,* Concrete, pp 35–37, May/June 1994

Appendix A – Authors' backgrounds and acknowledgements

Section	Authors' backgrounds and acknowledgements
Section 5.1 Calcium carbide moisture meter	**Chris Howard** is a chartered surveyor and Director of Consultancy for the Building Performance Unit (BPU), School of the Built Environment, Liverpool John Moores University. Chris Howard is grateful to **Graham Coleman**, dampness consultant, Bourton, Dorset, **David Cheetham**, Senior Fellow, School of Architecture & Building Engineering, University of Liverpool, **Dr Richard Clucas**, material scientist and **Michael Riley**, Head of Building Surveying Studies, Liverpool John Moores University for their supportive comments.
Section 5.2 Electrical capacitance meter	**Dr Kemal Ahmet (Timber)** is a physicist, currently leading the MORG (Moisture Research Group) at the University of Luton. Projects carried out by the MORG include moisture sensor development based on resistively measurement in hygroscopic materials. Studies into the possibility of developing capacitance type instruments for moisture gradient detection. Determination of the equilibrium moisture content of a range of timber species over a wide range of environmental conditions and laboratory tests on a wide range of moisture meters. **Andrew Tee (Roof and floor screeds)** is a director of Quest Technical Services Ltd specialising in the investigation, diagnosis and remedial repair of defects with the structure and fabric of buildings. He has considerable experience investigating moisture ingress problems associated with roofs, cladding and basements.
Section 5.3 Electrical earth leakage technique	**Stephen Thornton** is managing director of Thornton Consulting Group. He is a graduate of the Plastics and Rubber Institute (now incorporated into the Institute of Materials) and has been in the roofing industry since 1972, having been involved in many facets, including R&D, quality control, manufacture, sales and technical services. **Keith Roberts** is a director of Glanville Consultants, an independent firm of consulting engineers with a division specialising in roofing, cladding and building defects. He is a chartered civil and structural engineer with many years' experience in the inspection and reporting on the condition of roofing systems throughout the UK and Ireland. He is a regular contributor to specialist roofing and architectural journals and is co-chairman of an international CIB/RILEM Task Group developing the concept of "sustainable roofing".

Section	Authors' backgrounds and acknowledgements
Section 5.4 Electrical resistance meter	**Dr Kemal Ahmet (Timber)** For details, see **Section 5.2** Electrical capacitance meter. Dr Kemal Ahmet is grateful to **Guangya Dai**, **Sina Jazayeri** and **Richard Tomlin**, all from the Department of Technology, University of Luton, for help in compiling this section. **Tony Burke**, from the same department, provided many useful comments. **Chris Howard (Masonry and concrete)** For details, see **Section 5.1** Calcium carbide moisture meter.
Section 5.5 Humidity sensors	**David Mostyn (Mechanical hygrometers)** has been involved with the flooring industry since 1953. He is a member of the Institute of Chartered Arbitrators and the Expert Witness Institute and has carried out several cases as an arbiter, as well as numerous cases where he has been called as an "expert witness". **David Clifton (Electronic rh probes)** is the technical manager of Munters Ltd with over 25 years experience in the field of moisture mechanics and moisture control. The company, which operates internationally, is primarily involved in the identification and the control of moisture problems in buildings particularly moisture penetration in concrete floors, brick walls and timbers. **Prof. Jonathan G M Wood (Wooden plugs)** of Structural Studies & Design evolved and refined the Equilibriating Wooden Plug Method in the 1980s, with colleagues at Mott Macdonald Special Services Division, from his research work in the 1960s on the drying of crops and the control of deterioration of stored grains. The method has been used on a wide range of structures with AAR in the UK and overseas and has been used in determining in-situ rh in where high levels of moisture are creating conditions for rapid deterioration from corrosion and frost.
Section 5.6 Microwave moisture meter	**Jerry Assenheim** is a physicist/electronic and electrical engineer with forty years experience in the design of electronic measuring instruments. For the last 30 years has been managing director of Physical & Electronic Laboratories Ltd., a manufacturing consultancy producing precision electronic test equipment and solving industrial problems. For the last 20 years these have been primarily associated with moisture measurement in industrial applications, in particular with capacitance and microwave absorption techniques.
Section 5.7 Nuclear moisture gauge	**David Rolfe** is the technical manager for Controls Testing Equipment Ltd who is a major manufacturer of mainly laboratory based testing equipment for the construction industry. Over the past 15 years he has been responsible for providing a range of services for the portable nuclear moisture/density gauge user including training, applications advise, safety issues and all aspects of operation.

Section	Authors' backgrounds and acknowledgements
Section 5.8 Nuclear magnetic resonance	**Chris Sanders** joined Building Research Establishment's Scottish Laboratory in 1973 and has been involved with the analysis of dampness problems in buildings, both on building surfaces and within the fabric, and the development of guidance to avoid such problems. He has been concerned with the hygrothermal performance of building elements and has conducted site and laboratory tests, and developed models of moisture performance, including the use of NMR spectroscopy. He was a member of IEA Annex 24 on heat and moisture problems in the building envelope. He is also convenor of the CEN Committee developing European standards on moisture. He is actively involved with the BRE programme investigating the effects of climate change on future buildings.
Section 5.9 Oven drying method	**Chris Howard** For details, see **Section 5.1** Calcium carbide moisture meter.
Section 5.10 Radar	**Sam Dods** is the technical director of Aperio Limited. He graduated from Auckland University (NZ) Engineering Department in 1990 as a mining engineer. Since then he has worked in the field of non-destructive investigation employing a wide range of methods to investigate modern and historic structures in the UK and overseas. He has been involved in writing other guidance papers such as a document (to be published) for Historic Scotland on the use of non-destructive methods to investigate historic structures. The author is grateful to **Tony Goodier** of the BRE for his advice and contributions in preparing this section.
Section 5.11 Thermographic inspection	**George Ballard** worked in the fields of instrumentation and data interpretation at the Cambridge University Department of Geophysics, before founding GBG, a non-destructive investigation practice, applying new technologies to the engineering disciplines. His practice has been responsible for the introduction of a number of novel non-destructive methods, including both radar and thermographic inspection. He has been both co-editor and contributor to a wide range of technical publications on the use and application of non-destructive testing to engineered structures.
Section 5.12 Ultrasonics in the location of leaks	**Steve Pringle,** a partner in Messrs Sandberg, has been involved in the field of defect and failure investigations within the building industry for about 20 years. The investigation of water penetration into buildings has been a recurrent activity accounting for some 30 per cent of the departmental workload either through prevention in new-build or assessment and correction of active leaks.
Section 6. Automated long-term monitoring	**Tim Hutton** has been managing director at Hutton+Rostron Environmental Investigations for over 10 years. He is concerned with the performance and operation of buildings; investigation of building failures, particularly of moisture penetration into curtain walls, cladding and timber; the control of fungal and insect infestation of buildings; and the care and conservation of historic property.

Appendix B – Manufacturers and products

B1 TABLE LISTING MANUFACTURERS AND PRODUCTS

Manufacturer / Test method	Calcium carbide moisture	Electrical capacitance	Electrical earth leakage	Electrical resistance	Humidity sensors	Microwave moisture meter	Nuclear moisture gauge	Nuclear magnetic resonance	Oven drying	Radar	Thermographic inspection	Ultrasonics
American Gas and Chemical Co												♦
Ashworth Instrumentation	♦											
F Ball & Co					♦							
Brookhuis Micro-Electronics				♦								
Bruker Spectrospin								♦				
Campbell Pacific Nuclear							♦					
Cubbage Bullmann				♦								
Channel Electronics				♦	♦							
Delmhorst Instrument Co				♦								
Diplex					♦							
ERA Technology										♦		
Gann Mess-u Regeltechnik GmbH				♦								
General Eastern Instruments					♦							
Geophysical Survey Systems Inc										♦		
The Holt Trading Company					♦							
James Instruments Inc		♦										
Kernco Instruments Co. Inc												♦
K.P.M. Electronic Moisture Meters				♦								
Lignomat GmbH				♦								
MALA GeoScience										♦		
Physical & Electronic Laboratories						♦						
Protimeter		♦		♦	♦							
Riedel-de-Haon	♦											
Rotronic Instruments					♦							
Sensors & Software										♦		
Sovereign Chemical Industries		♦		♦								
Charles A Stewart					♦							
Testo					♦							
Tramex		♦		♦								
Troxler International Inc							♦					
Vaisala					♦							
Wagner Products Inc		♦										

B2 LIST OF MANUFACTURERS

American Gas and Chemical Co. Ltd.
220 Pegasus Avenue, Northvale New Jersey, 07647-1904, USA
Tel (00 1 01) 201 767 7300

Ashworth Instrumentation
Sycamore Avenue, Burnley BB12 6QR
Tel 01282 426554; fax 01282 438729

F Ball & Co Ltd
Churnetside Business Park, Station Road, Cheddleton, Leek, Staffs ST13 7RS
Tel 0153 8361633

Brookhuis Microelectronics,
PO Box 79, 7500 AB, Netherlands

Bruker Spectrospin Ltd.
Banner Lane, Coventry CV4 9GH
Tel 01203 855200

Campbell Pacific Nuclear *Contact:* ELE International Ltd
Eastman Way, Hemel Hempstead, Hertfordshire HP2 7HB
Tel 01442 218355; fax 01442 252472

Channel Electronics, Scientific Instrument Manufacturers
PO Box 58, Seaford, East Sussex BN25 3JE
Tel 01323 894961

Cubbage Bullmann Ltd
Harrem House, Ogilvie Road, High Wycombe, Buckinghamshire HP12 3DS
Tel 01494 441368

Delmhorst Instrument Co
PO Box 68, 51 Indian Lane East, Towaco, New Jersey 07082, USA
Tel (00 1 01) 973 334 2557

Diplex Ltd
133 St. Albans Rd, Watford, Herts WD1 1RA
Tel 01923 231784

ERA Technology Limited
Cleeve Road, Leatherthead, Surrey KT22 7SA
Tel 0372 374151; fax 0372 374496

Gann Mess-u Regeltechnik GmbH *Contact* Interwood Ltd
Woodworking Machinery, Stafford Avenue, Hornchurch, Essex RM11 2ER
Tel 01708 452591

General Eastern Instruments *Contact* ABLE Instruments & Controls Ltd
Cutbush Park, Danehill, Lower Earley, Reading, Berkshire RG6 4UT
Tel 0118 931 1188; fax 0118 931 2161; email dmachon@able.co.uk

Geophysical Survey Systems Inc *Contact* Allied Associates Geophysical Limited
17–19 Taylor Street, Luton, Beds LU2 0EY
Tel 01582 425079; fax 01582 480477

The Holt Trading Company Ltd
Long Furlong House, Holt, Norfolk NR25 7DD
Tel 01263 740370; fax 01263 741056

James Instruments Inc *Contact* Hammond Concrete Testing and Services Ltd
PO Box 75, Dorking, Surrey RH4 2YX
Tel 01306 887854; fax 01306 740433

Kernco Instruments Co Inc
420 Kenazo Avenue, El Paso, TX 79927 7338, USA
Tel (00 1 01) 915 852 3375

KPM Electronic Moisture Meters Ltd
150 Roundhay Rd, Leeds, West Yorkshire LS8 5LT
Tel 0113 249 6496

Lignomat GmbH *Contact* Scott & Sargeant Woodworking Machinery Ltd
Blatchford Road, Horsham, West Sussex RH13 5QZ
Tel 01403 273000

MALA GeoScience *Contact* Geodak Limited
16a Whitchurch Road, Pangbourne, Reading RG8 7BP
Tel 0118 9845380; fax 0118 9842531

Physical & Electronic Laboratories Ltd
28 Athenaeum Road, Whetstone, London N20 9AE
Tel 0181 445 7683

Protimeter plc
Meter House, Fieldhouse Lane, Marlow, Bucks SL7 1LX, UK
Tel 01628 472722; fax 01628 474312

Riedel-de-Haon *Contact* Laybond Products Ltd
The Wolff Tools Division, Riverside, Saltney, Chester CH4 8RS
Tel 01244 674774

Rotronic Instruments (UK) Ltd
Vector Point, Newton Road, Manor Royal, Crawley, West Sussex RH10 2TU
Tel 01293 571000; fax 01293 571008

Sensors & Software *Contact* Earth Science Systems Ltd
Unit 1, Kimpton Enterprise Park, Claggy Road, Kimpton, Herts SG4 8HP
Tel 01438 833611; fax 01438 833541

Sovereign Chemical Industries Ltd
Barrow in Furness, Cumbria LA14 4QU
Tel 01229 870800; fax 01229 870850

Charles A Stewart, Consultant Architect
Bellingdon Cottage, Bellingdon, Chesham, Bucks HP5 2XL
Tel 01494 785876

Testo Ltd
3 Oriel Court, Omega Park, Alton, Hampshire GU34 2QE
Tel 01420 544433; fax 01420 544434

Tramex Ltd
Shankill Businesss Centre, Shankill, Co. Dublin, Ireland
Tel 00 353 1 2823688; fax 00 353 1 2827880

Troxler International Inc *Contact* Controls Testing Equipment Ltd
Controls House, Icknield Way, Tring, Herts HP23 4JX
Tel 01442 828311; fax 01442 828466; email sales@controlstesting.co.uk

Vaisala UK
Suffolk House, Fordham Road, Newmarket, Suffolk CB8 7AA
Tel 01638 674400; fax 01638 674411

Wagner Electronic Products Inc
326 Pine Grove Road, Rogue River, Oregon 97537-9610, USA
Tel 00 1 541 582 0541; fax 00 1 541 582 4138